T0210750

Engineering Design

An Organic Approach to Solving Complex Problems in the Modern World

Synthesis Lectures on Engineering, Science, and Technology

Each book in the series is written by a well known expert in the field. Most titles cover subjects such as professional development, education, and study skills, as well as basic introductory undergraduate material and other topics appropriate for a broader and less technical audience. In addition, the series includes several titles written on very specific topics not covered elsewhere in the Synthesis Digital Library.

Engineering Design: An Organic Approach to Solving Complex Problems in the Modern World
George D. Catalano and Karen C. Catalano
2020

Case Studies in Forensic Physics
Gregory A. DiLisi and Richard A. Rarick
2020

An Introduction to Numerical Methods for the Physical Sciences
Colm T. Whelan
2020

Integrated Process Design and Operational Optimization via Multiparametric Programming
Baris Burnak, Nikolaos A. Diangelakis, and Efstratios N. Pistikopoulos
2020

Nanotechnology Past and Present
Deb Newberry
2020

Introduction to Engineering Research
Wendy C. Crone
2020

Empowering Professional Teaching in Engineering: Sustaining the Scholarship of Teaching
John Heywood
2018

The Human Side of Engineering
John Heywood
2017

Geometric Programming for Design Equation Development and Cost/Profit Optimization (with illustrative case study problems and solutions), Third Edition
Robert C. Creese
2016

Engineering Principles in Everyday Life for Non-Engineers
Saeed Benjamin Niku
2016

A, B, See... in 3D: A Workbook to Improve 3-D Visualization Skills
Dan G. Dimitriu
2015

The Captains of Energy: Systems Dynamics from an Energy Perspective
Vincent C. Prantil and Timothy Decker
2015

Lying by Approximation: The Truth about Finite Element Analysis
Vincent C. Prantil, Christopher Papadopoulos, and Paul D. Gessler
2013

Simplified Models for Assessing Heat and Mass Transfer in Evaporative Towers
Alessandra De Angelis, Onorio Saro, Giulio Lorenzini, Stefano D'Elia, and Marco Medici
2013

The Engineering Design Challenge: A Creative Process
Charles W. Dolan
2013

The Making of Green Engineers: Sustainable Development and the Hybrid Imagination
Andrew Jamison
2013

Crafting Your Research Future: A Guide to Successful Master's and Ph.D. Degrees in Science & Engineering
Charles X. Ling and Qiang Yang
2012

Engineering Design: An Organic Approach to Solving Complex Problems in the Modern World

George D. Catalano and Karen C. Catalano

ISBN: 978-3-031-00962-4   paperback
ISBN: 978-3-031-02090-2   ebook
ISBN: 978-3-031-00162-8   hardcover

DOI 10.1007/978-3-031-02090-2

A Publication in the Springer series
SYNTHESIS LECTURES ON ENGINEERING, SCIENCE, AND TECHNOLOGY

Lecture #13
Series ISSN
Print 2690-0300   Electronic 2690-0327

Engineering Design

An Organic Approach to Solving Complex Problems in the Modern World

George D. Catalano, Ph.D.
Department of Biomedical Engineering
Binghamton University

Karen C. Catalano, M.Ed.

SYNTHESIS LECTURES ON ENGINEERING, SCIENCE, AND TECHNOLOGY #13

ABSTRACT

While more and more undergraduate engineering programs are moving toward a multi-disciplinary capstone experience, there remains a need for a suitable textbook. The present text seeks to meet that need by providing a student friendly step by step template for this important and culminating academic journey beginning with the student design team's first meeting with the client to the final report and presentation. The text provides a wide range of design tools, a discussion of various design methodologies, a brief history of modern engineering, and a substantive consideration of engineering ethics. In addition, chapters are included on communication, team building and dealing with the inevitable obstacles that students encounter. Throughout the text, emphasis is placed upon the issues of environmental impact and the importance of diversity.

KEYWORDS

engineering design, capstone, design tools, design methodologies, communication, obstacles to success

Contents

CHAPTER 1

The Grand Journey

1.1 CAPSTONE DESIGN AS A JOURNEY

Capstone design is the culminating journey you as a student will take in the pursuit of your dream to be an engineer. That journey is very much the same as the hero's journey described by Joseph Campbell (Campbell, 2014 [16]). It is a cyclical journey that requires leaving the known world and entering the Unknown, and after encountering and overcoming many obstacles, returning to the known world with a story to tell (see Fig. 1.1).

Your journey is quite similar:

- You have mastered the basic engineering sciences and are feeling confident in your ability to pass any course.

- Suddenly, in capstone design, your Universe is totally upended, and you find yourself tossed into the uncertain world of design.

- You go out in search of a solution, overcoming a wide array of obstacles along the way.

- Then you return and share your knowledge, your information, and ultimately your final design.

1.2 YOUR GUIDEBOOK FOR THE JOURNEY

This text is a guide for any engineering design team confronting complex engineering design problems in the 21st century. It outlines a path from the first meeting with the project's client to the final presentation of the completed project to client and professor. We are offering a new design approach to today's complex engineering design challenges with a deeply held commitment to the health of the Earth, its cultural diversity, and its diminishing numbers of living species.

In addition to a step-by-step approach to the various stages of design (Fig. 1.2), we have included:

- critical and creative thinking skills necessary to successfully complete a capstone design project;

- some historical context for your design work;

- ways to improve team dynamics;

Figure 1.1: Joseph Campbell's hero's journey.

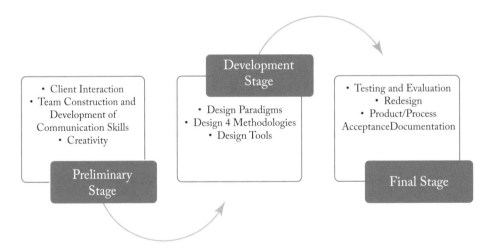

Figure 1.2: Flow process for engineering design.

- a variety of design methodologies including traditional design, concurrent design, green design, eco-efficient design, eco-effective design, and others;

- a wide array of design tools including the client statement, the statement of work, TRIZ, pairwise comparison, function means, rich pictures, and others;

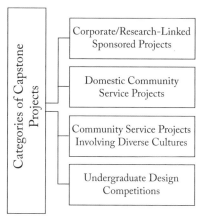

Figure 1.3: Categories of capstone design projects.

- emerging new ideas in the design process;

- an overview of some of the complex ethical challenges engineers face today;

- insights into the development of testing, evaluation, and follow-through of redesign stages;

- obstacles to expect on your journey; and

- how to write and present your final report.

There are four broad categories of capstone design projects that you may encounter at the undergraduate level (Fig. 1.3):

(1) corporate/industrial/research-linked sponsored projects,

(2) domestic community service projects,

(3) community service projects involving diverse cultures, and

(4) undergraduate design competitions.

This text is a guidebook to the skills a design team needs to successfully meet the challenge presented by any of these categories.

Our Back Page: Why this book?

This text is the product of nearly a half century of experience in the teaching of capstone engineering design. That experience was garnered at both large research—focused universities as well as small undergraduate colleges and academies. We have taught capstone design in the contexts of one engineering discipline (mechanical and aerospace engineering) as well as multi-disciplines (mechanical, electrical, computer, aerospace, systems, industrial, bioengineering, and biomedical). We also have worked with non-engineering disciplines including business, management, sociology, education, and the fine arts. We have witnessed the increasing need for capstone design to become more multi-disciplinary as engineering is confronted by ever more complicated challenges.

Throughout these nearly 50 years, we have not encountered a text which adequately addresses the needs students have in fulfilling their capstone design requirements. Our hope is that the present text will provide an approach for both students and faculty alike to elevate the quality of their design experiences to the highest possible level. We also wish to support efforts in engineering that address the complex challenges in the 21st century in ways which seek to make the world a kinder, gentler place.

1.3   THE MOST CHALLENGING UNDERGRADUATE COURSE

Many undergraduate engineering students report that capstone design is the most challenging course in their engineering education. Although there are many reasons for this, probably the most important one is that design requires the most advanced critical thinking skills.

In 1956, Bloom with collaborators published a framework for categorizing educational goals called the "Taxonomy of Educational Objectives" (Anderson, 2000 [4]). Notice that "creating" (i.e., design of something that did not exist before) rests at the very highest level of the pyramid (Fig. 1.4).

As engineering students progress from calculus and physics to more advanced courses, they begin with simple recall of facts from memory and hopefully culminate with a sufficiently deep understanding that allows them to apply their knowledge and analyze problems. Next comes evaluation (using knowledge to make decisions) and finally creating (designing and constructing a product and/or process).

Design is challenging because:

- it requires the engineer to solve a problem that has never been solved before. There are no approved solutions;

- it requires creativity in addition to fundamental engineering knowledge. Design is synonymous is creation;

- it requires students to work effectively with other people—in teams as well as with clients—some of which may have completely different backgrounds and/or different cultures; and

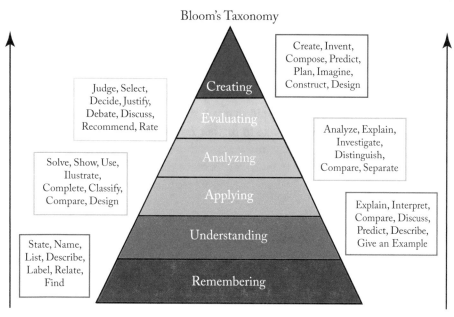

Figure 1.4: Bloom's taxonomy (information learning activity—results and analysis, 2013 [40]).

- most importantly, it demands both the courage to offer new solutions and the humility to know when a proposed solution is not appropriate.

1.4   SELECTING A PROJECT

After several years of hard work as an undergraduate, the time arrives for you to choose a design project of special interest—one that will test all that you have learned during your undergraduate career. Here are six tips to guide you as choose your capstone project (Varsity Tutors, 2015 [67]).

1. **Brainstorm**

   By the time you are ready to begin thinking about your capstone project, you have likely fulfilled most of your degree requirements. Before you begin researching your project, take the time to reflect.

   - What topics or classes were of interest to you?
   - Are there any topics that you would like to learn more about?

2. **Read**

   Once you have a broad list of potential topics, preview the latest developments in these areas.

- Does any research look particularly promising?
- Can you see yourself contributing to the discussion surrounding any of these topics?

3. **Narrow your focus**

   At this stage, try to identify one topic that you would like to further explore, and develop several questions that you would like to further research and possibly expand upon. Write down everything that you already know about these topics and hypothesize about how a proposed project might be able to address aspects of the topic that have yet to be developed.

4. **Consult with your project adviser**

   Your advisor will be able to redirect your focus if necessary, as well as recommend resources that can further your understanding of the topic.

5. **Read more**

   By now, you should have a specific project in mind as well as questions that you will explore in your capstone project. Do more research to gather more information surrounding this project idea.

6. **Begin work!**

   Taking these simple steps in preparation for your capstone project can help to ensure that you choose an interesting and challenging project to explore as a summation of your college education.

1.5   CAPSTONE PROJECTS FROM THE PAST

We have selected a variety of actual student capstone projects to help illustrate various stages of the design process as it unfolds. You will see them highlighted throughout the text.

- Project 1: **Waste-to-Wealth.** This project focused upon the design and delivery of a device that transforms plastic trash into usable sheets of plastic material. The design process began with the team meeting with the client and discussing the waste issue occurring within the rural community of El Charcón, El Salvador (see Fig. 1.5). Through this discussion, the team learned that, because the village is the last village on the river before it feeds into open waters, the village was overwhelmed with large amounts of commercial waste flowing down the river from higher villages. Trash was accumulating and littering the town as the river flowed down through their river and into the ocean.

  The team was tasked with creating a device or process to transform this waste into a valuable resource. Upon investigating the situation, the team discovered that the most prominent wastes in the area are plastics, specifically thin film plastics like bags and

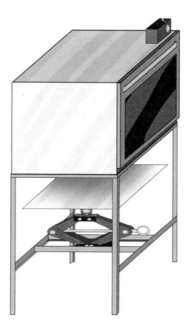

Figure 1.5: Waste to wealth design project.

wrappers. It was hypothesized that most of these wastes were type 2 plastics, high-density polyethylene (HDPE) or polyethylene, which comprise a vast majority of plastic bags, containers, and bottles thrown into landfills each year.

- Project 2: **ALS Enhanced Mobility Device.** This project sought to develop a device to help their client who suffers from Amyotrophic Lateral Sclerosis (ALS), a progressive neurodegenerative disorder that takes away a person's ability to control basic muscular movements due to motor neuron degeneration. Because the mechanisms behind its onset are not well understood there is currently no cure or treatment for ALS.

    As a result of this disease, their client, an accomplished artist, had limited control of her upper body and subsequently lost the ability to perform activities she is passionate about, including painting, dancing, and puppetry. The objective of this project was to create a device that held the client in an upright position so that she could stand, move freely, dance without additional assistance, and use her feet rather than her hands to paint. The ALS Enhanced Mobility Device was designed to increase the independence of the client and allow her to perform some of the activities she used to enjoy (see Fig. 1.6).

- Project 3: **Body-on-a-Chip.** The low efficiency of drug success in clinical trials indicates that there is a lack of preclinical models capable of providing accurate predictions

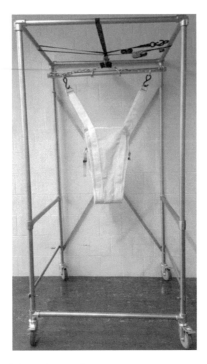

Figure 1.6: **ALS** enhanced mobility device design project.

of human responses to novel drugs. "Body-on-a-chip" systems attempt to mimic tissue-tissue interactions by connecting fluidic streams from several different types of tissue cultures so that metabolites are produced, consumed, and exchanged in physiologically relevant concentrations.

The team designed, constructed, tested, and characterized a body-on-a-chip using a reusable 3D-printed mold to imprint compartments and channels into a polymer-based scaffold (see Fig. 1.7). Their design incorporated five different organs relevant to drug metabolism and bioavailability onto the body-on-a-chip as well as compartments and channels created by scaling values found in the human body. The design team created a linear flow pattern modeled after physiological flow and then characterized the flow rates, residence times, fluid to tissue ratio, and cell viability within their device.

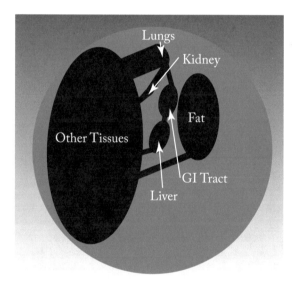

Figure 1.7: Body-on-a-chip design project.

CHAPTER 2

The Grand Challenges

"The more complex the world becomes, the more creative we need to be to meet its challenges."

Sir Ken Robinson (Robinson, 2014 [59])

2.1 CHALLENGES FACING ENGINEERING IN THE 21ST CENTURY

Since the beginning of the 20th century, engineering has evolved in ways far beyond the limits of human imagination. As a profession, engineering has been primarily responsible for the development and distribution of electricity, the automobile, the airplane, clean water supply and distribution, electronics, radio and television, agricultural mechanization, computers, the telephone, air-conditioning and refrigeration, the Internet, imaging technologies, household appliances, vast highways, spacecraft, petroleum and petrochemical technologies, health technologies, lasers and fiber optics, nuclear technologies and high-performance materials…to name just a few! The future promises even more possibilities for tomorrow's engineers. However, with the excitement that accompanies advances in technologies, new, and ever more complex challenges also arise.

In 2017 the National Academy of Engineers (NAE) wrote that while "engineering has revolutionized and improved virtually every aspect of human life, …the century in which we now find ourselves poses challenges as daunting and formidable as any that humankind has ever faced in the past" (National Academy of Engineers, 2017 [51]). That same year the NAE convened a select international committee to evaluate ideas on the greatest challenges and opportunities for engineering. The Academy referred to these challenges as "Grand Challenges" and the list included making solar energy affordable, providing energy from fusion, developing carbon sequestration methods, managing the nitrogen cycle, providing access to clean water, restoring and improving urban infrastructure, advancing health informatics, engineering better medicines, reverse engineering the brain, preventing nuclear terror, enhancing virtual reality, advancing personalized learning and engineering new tools for scientific discovery. In today's vernacular, these "grand challenges" are often referred to as *complex problems*, ones that require an interdisciplinary approach from multiple disciplines.

Our Back Pages: Simple, Complicated, and Complex Systems
Consider the following classifications:

- A simple system is one that has one path that leads to one result.

- A complicated system is characterized by multiple paths leading to one result.

- A complex system is one that has multiple paths that lead to multiple results.

Examples of complex systems include Earth's global climate, organisms, the human brain, infrastructure (such as power grid, transportation or communication systems), social and economic organizations (like cities), an ecosystem, a living cell, and ultimately the entire universe. Trying to model and understand the behavior of such systems is incredibly challenging yet more and more engineers are expected to meet these challenges.

2.2   COMPLEX ENGINEERING DESIGN PROBLEMS

A complex engineering design problem includes one or more of the following characteristics:

- wide ranging or conflicting technical issues having no obvious solution;

- problems not encompassed by current standards and codes;

- diverse groups of stakeholders;

- many component parts or sub-problems;

- multiple disciplines necessary to address the problem; and

- significant consequences in a wide range of contexts.

Notice several important elements.

1. Issues associated with any design may be both wide ranging and conflicting.

2. Because the problems are unique and open-ended, engineers must have both critical and creative thinking skills on-demand.

3. Individuals involved in the design problem may be (and often are) widely diverse requiring a sensitivity to a wide range of cultures.

4. The expertise of multiple disciplines is essential—multiple engineering disciplines at a minimum—but also possibly the inclusion of disciplines outside engineering and the sciences.

An example of a complex problem involves the sustainability of modern society:

How do we as engineers focus on meeting the needs of the present without compromising the ability of future generations to meet their needs?

The challenge presented by sustainability includes multiple design paths that can result in multiple end results.

2.3   A NEW METHODOLOGY: AN ORGANIC APPROACH

Traditionally, engineering has used strictly analytical tools to arrive at solutions to design problems. While this strictly analytical approach has worked well in the past, the challenges created by increasingly more complex problems require a new methodology.

We believe that an effective design strategy for addressing complex problems requires both an analytical approach as well as a holistic systems approach. In this text we are calling this new methodology an "organic" approach. Borrowed from the study of Earth's ecosystems, the term "organic" denotes an association or relationship among the parts of the whole which allows those individual parts to fit together harmoniously.

For example, in an ecosystem, the individual is a member of a species which in turn is in relationship to countless other species. The health of an eco-system depends not only upon the health of individuals and individual species; it also depends on the health of relationships among the different individuals and species. A healthy ecosystem is one that is sustainable; that means it can maintain its structure (organization) and function (strength) over time in the face of external stress (resilience).

Consider this organic approach in nature as illustrated in Fig. 2.1.

An individual species—in this case the monarch butterfly—exists in a population of individuals of the same species which, in turn, exists in an ecosystem that includes both predators and prey species for the butterflies as well as many other species.

Using this same organic approach, we discover that a proposed sustainable engineering design solution is similarly embedded in a local/global societal context as well as the local/global environment, as shown in Fig. 2.2.

Our Back Pages: Why an expanded "organic" view is needed

The need for moving away from strictly an analytical approach to a holistic systems approach can be illustrated using the predator-prey problem, a classic problem in modeling populations.

In the study of the dynamics of a single population, we typically take into consideration such factors as the "natural" growth rate and the "carrying capacity" of the environment. In this classic problem we have two populations that interact with each other—a predator and its prey. To keep the model simple, the following assumptions are made: the predator species is totally dependent on a single prey species as its only food supply; the prey species has an unlimited food supply; and there is no threat to the prey other than the specific

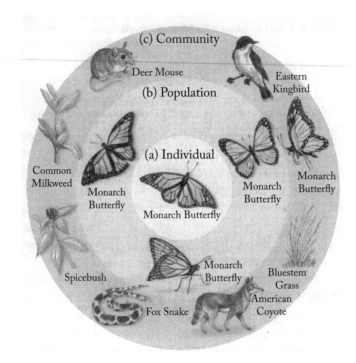

Figure 2.1: Organic understanding of the natural world (Rissignol, 2012 [58]).

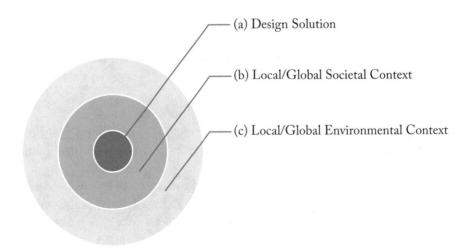

Figure 2.2: Organic engineering design.

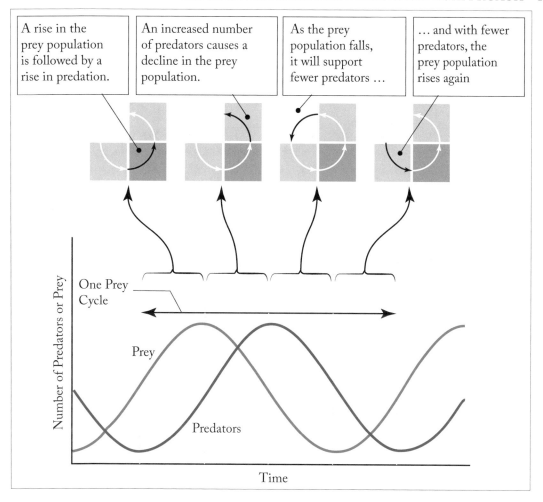

Figure 2.3: Classical predator prey problem (Ricklef, 2015 [57]).

predator. Classical differential equations are used to model the wolf-rabbit interaction with typical results are shown in Fig. 2.3. Note the behavior is cyclical and continues indefinitely.

Another interesting example with historical data is the wolf-moose interaction in Isle Royale in Lake Michigan. Real data for the actual wolf-moose population curves is shown in Fig. 2.4.

Clearly, the actual population numbers behave very differently. What might be contributing reasons for the discrepancies?

While models have their place in advancing our understanding of the natural world, they are limited. In the wolf-moose model, for example, there is no mention of packs and pack structure, nor room for individuality or any emotions. The same observation is valid for the moose. In addition, countless other species in the local ecosystem who may have had significant effect on the numbers are not included nor are the attitudes of local residents to wolf presence.

In the past, science and engineering judged nature and natural systems as simply a collection of objects—much like the gears and levers in a mechanical clock. They are not! There certainly is much more to be considered. An organic approach to such a complex problem insists that not only the individual parts—the moose or the wolves—be considered but also their interactions among their own species as well as other species.

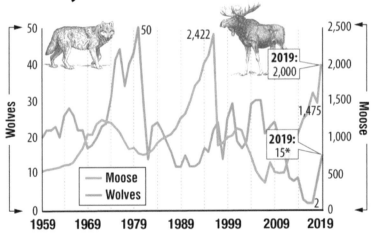

Figure 2.4: Isle Royale moose-wolf interactions (Myers, 2019 [50]).

2.4   QUESTIONS FOR THOUGHTFUL CONTEMPLATION

1. Define a simple, a complicated, and a complex problem. Give an example of each and discuss how you arrived at that determination.

2. The predator-prey problem brings up issues related to how we imagine the natural world. Do you see it as strictly mechanical—analogous to collections of inanimate objects—or is it something else? Discuss how you "imagine" the natural world.

CHAPTER 3

Revolutions in Modern Engineering

All truths are easy to understand once they are discovered; the point is to discover them.

Galileo Galilei (Galilei, 2017 [37])

3.1 THE IDEA OF TECHNOLOGICAL REVOLUTIONS

A technological "revolution" is a period in which one or more technologies is replaced by another technology in a short amount of time. It is an era of accelerated technological progress characterized by new innovations whose rapid application and diffusion cause an abrupt change in society. Technological revolutions are significant because they shape the future of social and cultural development as evidenced by the scientific, industrial, the information technology, and the synthetic biology revolutions.

Our Back Pages: Revolutions
> The term "revolution" ultimately goes back to the Latin *revolvere* "to revolve, roll back." When the term "revolution" first appeared in English in the 14th century, it referred to the movement of a celestial body in orbit. In Ptolemy's model of the heavens, the Earth was viewed as a stationary body at the center of the Universe with the planets and stars revolving around it. Copernicus changed all that—the Sun was stationary and the Earth "revolved" around the Sun. At the time Copernicus's idea was very controversial. Although Copernicus was careful to declare his findings as theory, as theoretical, nevertheless, it was the start of a change in the way the world was viewed, and Copernicus came to be seen as the initiator of the Scientific Revolution.

3.2 REVOLUTIONS IN TECHNOLOGY IN THE LAST 1000 YEARS

Engineering has played an integral part in the six technologically driven revolutions that have occurred over the course of the last seven centuries (see Fig. 3.1). They include:

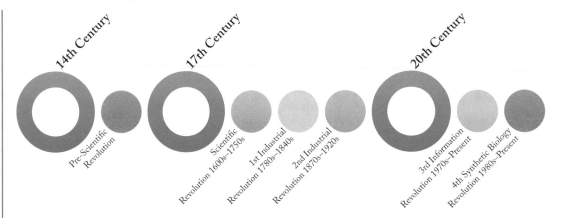

Figure 3.1: Timeline for technological revolutions.

1. Pre-Scientific Revolution (1300s–1600s).

2. Scientific Revolution (1600s–1750s).

3. First Industrial Revolution (1780s–1840s).

4. Second Industrial Revolution (1870s–1920s).

5. Third Industrial/ Information Revolution (1970s–present).

6. Fourth Industrial/Synthetic Biology Revolution (1980s–present).

3.3   THE PRE-SCIENTIFIC REVOLUTION (1300s–1600s)

The Renaissance has traditionally been recognized as a period of an extraordinary flowering of arts and letters. The persistence of this view has long obscured the revival of technical activity that began in the late 14th century—notably in Italy—and lasted, with undiminished vigor, through the 18th century. A closer examination of the period shows that the prime movers in this pre-scientific revolution were, in most cases, the very same "artists" ("artificers" might be a more suitable term) who led the radical renewal of painting, sculpture, and architecture during those decades.

Renaissance artists were routinely involved in activities that we would now define as engineering. Moreover, their workplace—the celebrated "artist's workshop" of the Renaissance—had more in common with a bustling factory than with the modern conception of an artist's studio.

These forerunners of engineers—the practical artists and craftsmen—proceeded mainly by trial and error. Tinkering combined with imagination created many marvelous devices including the printing press, double shell domes, the crank and connecting rod, the parachute,

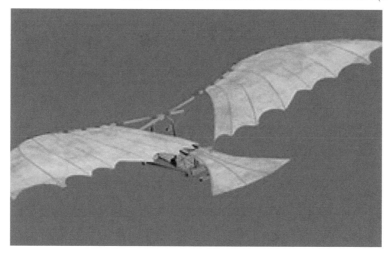

Figure 3.2: Leonardo daVinci's flying machine (DaVinci Inventions—Inspired Engineering, 2014 [25]).

the mariner's astrolabe, and the floating dock to name a few. It was during this time that the term "engineer" was first used as these individuals were seen to possess "ingenuity" and were considered "geniuses."

Leonardo daVinci is the most famous representative of this time; he was a brilliant scientist and engineer as well as a peerless artist. His greatest inventions are regarded as the product of a towering genius who anticipated by many centuries the technical achievements of the modern age. One of his designs that anticipated flight is shown in Fig. 3.2.

While studying birds in flight, da Vinci realized that humans would not be strong enough to fly if they used only wings attached their arms due to their excessive weight. He envisaged that to achieve flight an apparatus would require include levers, pedals, and pulleys. On this basis, in about 1490 Leonardo da Vinci drew his up plans for a flying machine that would keep a man in the air by the beating of its wings. If you look at the plan in the drawing, it shows a pair of giant wings that connect to a wooden frame. The pilot would lie face down inside the frame on a board. Using his hands, the pilot would grip a stick coming down from each wing for direction control. Because there was no engine, to achieve flight, the pilot would make a flapping motion by pushing his legs downwards with his feet held in two spurs. While the design was strictly theoretical, daVinci anticipated modern aerodynamics.

Leonardo daVinci was not the only "genius" who embraced art, science, and technology; he had many contemporaries. Among the other great artists-engineers of the Italian Renaissance were Filippo Brunelleschi, Mariano di Iacopo, and Francesco di Giorgio (Bjerklie, 1998 [11]). The "mechanical marvels" they created include a three-speed hoist driven by oxen and capable of lifting loads of more than a ton to a height of 270 feet, a revolving crane, a mobile siege

Figure 3.3: Florence Cathedral of Brunelleschi (wordpress.com, 2013 [72]).

bridge, and a paddleboat, as well as such architectural masterpieces as the dome of the cathedral of Florence, 100 feet high and 165 feet across—a daring piece of engineering and construction that remains an enduring symbol of the Renaissance (Fig. 3.3). In fact, this period in Italy has been called by some historians the "renaissance of machines," a phrase still unfamiliar today because it has been dwarfed and obscured by the revival of arts and letters.

3.4   THE SCIENTIFIC REVOLUTION (1600s–1750)

The Scientific Revolution was a series of events that marked the emergence of modern science. In this revolution, developments in mathematics, physics, astronomy, biology (including human anatomy), and chemistry transformed the West's views of nature. This revolution took place in Europe toward the end of the Renaissance period and continued through the late 18th century, influencing the subsequent intellectual and social movements which together are known as the Enlightenment. While the actual dates are debated, the publication in 1543 of Copernicus's *On the Revolutions of the Heavenly Spheres* (Copernicus, 1995 [22]) is often cited as its beginning with the publication of Galileo's *Two New Sciences* (2017 [37]) in 1632 signaling its conclusion.

Copernicus was a Polish astronomer known as the father of modern astronomy. He was the first modern European scientist to propose that Earth and other planets revolve around the Sun. Prior to the publication of his major astronomical work, European astronomers argued that Earth lay at the center of the universe, the view also held by most ancient philosophers and biblical writers. In addition to correctly postulating the order of the known planets (including Earth) from the Sun, and estimating their orbital periods relatively accurately, Copernicus argued that Earth turned daily on its axis and that gradual shifts of this axis accounted for the changing seasons.

Galileo's *Two New Sciences* sought systematic explanations and adopted a scientific approach to practical problems. It is a landmark regarded by many engineering historians as the beginning of the scientific method. In presenting his findings, Galileo formulated a methodology for rigorous experimental study of natural phenomena that became the foundation for modern experimental science.

This period's completion is attributed to the "grand synthesis" that was achieved by Isaac Newton's *Principia*, published in 1687 (Newton, 2016 [52]). Newton laid out in mathematical terms the principles of time, force, and motion that have guided the development of modern physical science. Even after more than three centuries and the theories of Einsteinian relativity and quantum mechanics, Newtonian physics continues to account for many of the phenomena of the observed world.

This phase of scientific exploration led to the First Industrial Revolution, when machines, increasingly powered by steam engines, started to replace muscles in most activities.

Our Back Pages: Galileo

Galileo Galilei was a Tuscan (Italian) physicist, mathematician, astronomer, and philosopher who played a major role in the Scientific Revolution. His achievements include improvements to the telescope and consequent astronomical observations, as well as support for Copernicanism. Galileo has been called the "father of modern observational astronomy," the "father of modern physics," the "father of science," and "the father of modern science."

The motion of uniformly accelerated objects, included in nearly all high school and introductory college physics courses, was studied by Galileo as the subject called "kinematics" (see Fig. 3.4).

His contributions to observational astronomy include the telescopic confirmation of the phases of Venus, the discovery of the four largest satellites of Jupiter (which were named the "Galilean moons" in his honor), and the observation and analysis of sunspots. His work in applied science and technology also improved the design of the compass.

Figure 3.4: Kinematics and projectile motion.

3.5   FIRST INDUSTRIAL REVOLUTION (1780–1840)

From the late 1800s through the early 19th century, Western society's view of the work of engineers underwent a dramatic shift. Where once viewed as practical artists, engineers began to be perceived as scientific professionals. The French, who were at the forefront civil engineering with an emphasis on mathematics, developed university engineering education under the sponsorship of the French government while the British pioneered mechanical engineering and autonomous professional societies under the laissez-faire attitude of their government. Gradually, practical thinking became not only intuitive but more and more scientific as engineers developed mathematical analysis and controlled experiments.

While the mechanical clock frequently serves as a metaphor for the Scientific Revolution, the factory symbolizes the First Industrial Revolution. After patenting his spinning frame in 1769, the British engineer Richard Arkwright created the first true factory at Cromford, near Derby (The Editors of Encyclopedia Britannica [62]). Nothing had ever been seen like this before, and it changed not only Great Britain but also the world.

Prior to this factory, the original cottage process for spinning fabric involved two or three people working in their own home. In a few years after its opening creation, Arkwright's factory employed over 300 people, and by 1789, the mill employed 800 people. Except for a few engineers in the factory, the bulk of the work force was essentially unskilled. They had their own tasks to do over a set number of hours. Whereas those in the cottage system could work their own hours and enjoyed a degree of flexibility, those in the factories were governed by a clock and factory rules. Children were frequently employed in the factories that sprang into existence for the very sad reasons that there were plenty of them in orphanages, they could be replaced easily if accidents did occur, and they were much cheaper than adults.

Figure 3.5: Steam engine (Palermo, 2014 [54]).

Figure 3.6: Steam locomotive (Technical Revolution, 2010 [61]).

Our Back Pages: Steam Engines and Locomotives (Bellis, 2020 [9])

Modern civilization was forged in the factories of the Industrial Revolution powered by the steam engine (Fig. 3.5). The first crude steam-powered machine, built by Thomas Savery in Britain in 1698, was designed to pump water out of coal mines. In 1712, Thomas Newcomen built a machine where the steam pushed a movable piston in one direction. In 1763, James Watt designed a reciprocating piston-in-cylinder arrangement and, more importantly, he found out a way to make this back-and-forth motion turn a wheel using a crankshaft. This advance ultimately led to the steam locomotive (Fig. 3.6).

3.6   SECOND INDUSTRIAL REVOLUTION (1870–1920)

While technology has changed the world in many ways, perhaps no period introduced more changes than the Second Industrial Revolution. Throughout the Western world from the late 19th to early 20th centuries, cities grew, factories sprawled, and people's lives became regulated by the clock rather than the sun. Rapid advances in the creation of steel, chemicals, and electricity helped fuel production of mass-produced consumer goods and weapons, and travel became much easier due to the expansion of railroad and the introduction of the automobile. Simultaneously, ideas and news spread rapidly via newspapers, the radio, and telegraph. Life greatly accelerated its pace.

This was an era in America when industrial growth created a class of wealthy entrepreneurs and a comfortable middle class supported by workers made up of immigrants as well as arrivals from America's farms and small towns. Many workers came to industries from rural backgrounds where their work was self-directed and organized around the seasons and changing light. During this period, many found themselves working in a factory that was clock-regulated and unchanging through the seasons.

The Second Industrial Revolution also took place at a time of rapid territorial expansion in America. Gazing out across vast stretches of land and possessing a sense of "manifest destiny," American entrepreneurs rushed to extract raw materials such as ore, coal, and iron from the abundant natural resources that land offered. Railroads provided the infrastructure for transport, and there was a huge expansion of communication networks because of the invention of the telegraph.

Our Back Pages: Chicago World's Fair 1893 (Davis, 2016 [26])

On May 1, 1893, the Chicago World's Fair opened to the public. Up until its closing ceremonies in October 1893, it gave fairgoers the opportunity to observe the latest groundbreaking inventions. Today these items, which were met with both wonder and enthusiasm, blend into the landscape of our modern world. Some of the more notable inventions include the zipper, spray paint, and dishwashers—not to mention Wrigley's Juicy Fruit gum, Cracker Jacks, and Pabst Blue Ribbon beer!

Also unveiled at the Fair was the original Ferris Wheel (Fig. 3.7). This huge structure was supported on two 140-foot towers and a 45-foot axle and had a diameter of 250 feet and a height of 264 feet. It included cars that were 24 feet long and 13 feet wide, each carrying up to 60 passengers. A single revolution of the great wheel took 9 min.

Figure 3.7: First ferris wheel at 1893 Chicago World's fair (Larson, 2004 [45]).

3.7   THIRD INDUSTRIAL (INFORMATION) REVOLUTION (1970–PRESENT)

While the First Industrial Revolution supplanted cottage industries with factories and the Second Revolution made mass production by assembly lines the norm for industrial manufacturing, the Third Industrial Revolution involves information. Sometimes called the "Age of Information," it refers to the period from the 1970s onward and is characterized by unprecedented advancements in technical innovation as well as the use of new digital information and communication technologies in everyday life.

In this revolution there was a rapid shift from an economy based on industry to an economy based upon information technology. It began with the development of transistors which led to advances in microminiaturization in computing. Digital technology soon replaced analog technology, and the workforce became globalized.

Advances in computers, computerized machinery, fiber optics, communication satellites, and the internet combine to drive ongoing advances in robots and 3-D printing (Fig. 3.8). Currently, robots are isolated from people for security reasons; their unthinking motions are a safety hazard. This next generation of robots will not only be cheaper and easier to use but also be able to work with people rather than replace them. Future robots will be tuned to serve human workers by fetching and carrying parts, holding and sorting items, cleaning, sanitizing, and countless other tasks. 3-D printing—more accurately termed additive manufacturing—is changing the way products are designed and fabricated. More and more work is being done in front of computer screens and less in factory assembly lines (Fig. 3.9). This allows companies to quickly adjust to changes in local tastes.

Figure 3.8: SLO camera made with 3-D printing (Dudley, 2016 [29]).

Figure 3.9: Turbulent flow simulation with 3-D Printing (Thurow, 2009 [64]).

3.8   FOURTH INDUSTRIAL (SYNTHETIC BIOLOGY) REVOLUTION (1984–PRESENT)

Synthetic biology is a multidisciplinary area of research using enzymes, genetic circuits, and cells not only to create new biological parts, devices, and systems but also to redesign systems that are already found in nature. It involves applying engineering principles to biological systems.

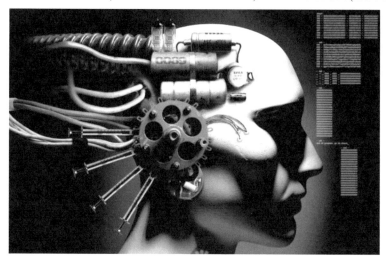

Figure 3.10: Transhumanism (Anthony, 2013 [5]).

Synthetic biology began in 1984 when Harvard professor Steven Benner synthesized a gene for an enzyme for the first time. The emergence of synthetic biology has revolutionized industrial biotechnology. Using genetic engineering, biotechnology manipulates the enzymes and genes of living organisms or their components to produce commercial products such as pest resistant crops, new bacterial strains, or novel pharmaceuticals. Examples of biotechnology today include:

- genetic crop manipulation;

- animal cloning;

- human gene therapy;

- alternative sources of fuel; and

- the development of transhumans (Fig. 3.10).

Our Back Pages: A Brave New World

Aldous Huxley's *Brave New World* (1932 [39]) takes place in a futuristic society in which people are grown as nearly identical embryos in bottles and conditioned to remove not only strong desires and emotions, but also the need for human relationships and strong emotions. Members of society take a drug called soma to help keep them docile. Huxley's work anticipated huge advances in reproductive technology, psychological manipulation, and classical conditioning. The reality today is that we are rapidly approaching many of the conditions which Huxley wrote about almost 100 years ago.

Advances in synthetic biology and biotechnology raise an incredible number of ethical questions that future engineers will confront (Frischman, 2018 [36]). Today biotechnology engineers and researchers must address five core concerns that are, in fact, the foundations of bioethics.

Concern 1. Use of genetics

While new personalized medications can eliminate side effects, the use of genetic information to create medicine contributes to the rising cost of drugs and shifts attention away from designing affordable drugs available for mass production. Another critical bioethical issue for geneticists is the creation of designer babies. Parents will soon be able to make a host of changes in the gene makeup of their offspring. The ethical questions raised by this are limitless.

Concern 2. Use of stem cells

The use of stem cells is one of the most controversial issues in biotechnology, especially in the United States. Supporters argue that the embryos used to generate stem cell lines in cancer research have the potential to save untold numbers of lives. Opponents believe the destruction of any embryo for research purposes is an ethical violation.

Concern 3. Human testing

Science has made many fascinating advances in medicine. This has all been possible with numerous trials and experimentations. Human experimentation is an essential step to test effectiveness of a chemical or technique to help develop effective treatments for individuals. Questions arise when considering "just" treatment and respect for all human life. Also, informed consent of the test subject is a particularly important ethical concern. The Tuskegee Syphilis Study, cruel Nazi experimentations in German camps, and the Willowbrook School Hepatitis study are just some examples of atrocities done on humans in experimentation with no regards to morality or ethical consideration.

Concern 4. GMO food

A growing global population, concerns about diminished drinking water reserves, and a need to keep food affordable all prompted the expansion of the availability of genetically modified organisms (GMOs) in the food supply. Opponents of GMOs claim the organisms place the entire food supply at risk through the homogenization of plant life and the death of biodiversity. They also argue that insects and plant-destroying bacteria or diseases will continue to evolve with the GMOs, resulting in super-pests and super-diseases that are untreatable by modern methods.

Concern 5. Costs

Developing new drugs and instruments that improve the quality of life and extend life expectancy to new highs are all lofty goals, but they come at a significant financial cost. This is particularly true in the United States. Questions of access and availability will become more urgent in the future as the costs continue to rise.

3.9   QUESTIONS FOR THOUGHTFUL CONTEMPLATION

- Name and describe the six revolutions discussed in this text. How are they similar? How are they different?

- Identify one key individual and one key advance made in each revolution. Explain why you selected each.

CHAPTER 4

The Current State of Engineering

Come gather 'round, people wherever you roam
And admit that the waters around you have grown
And accept it that soon you'll be drenched to the bone
If your time to you is worth savin'
And you better start swimmin' or you'll sink like a stone
For the times they are a-changin'

Bob Dylan (Dylan, 1964 [30])

4.1 EXPANDING PERSPECTIVES: THE RIPPLE EFFECT

If you have ever dropped a pebble in a pool of still water, you watched as the ripple surrounds the impact location and then gradually moves outward in concentric waves (Fig. 4.1). As we explore traditional modern engineering, green engineering, and other emerging trends in engineering, we can use this image to visualize how emerging branches are extending the boundaries of responsibilities in the engineering profession.

4.2 TRADITIONAL ENGINEERING

Traditional engineering is based on decomposition—taking a complex system and breaking it into a set of smaller, less complex systems. To understand the whole of something, you break it into the parts, analyze the parts, and put them back together again. A good example of traditional engineering is an automobile. You can break the car down into its various components such as the chassis, the body, the steering, the transmission, the engine, the brakes, and other sub-systems. Once you analyze each of these sub-systems, you will have developed a greater understanding of the whole of the car.

Traditional engineering relies mainly on a sequential process in which the various tasks involved in the design and manufacturing of a product are performed in a pre-defined and set order. The focus is on the end-product. While this approach has many strengths, it can lead to certain drawbacks such as a loss of flexibility in the entire process and large design modifications

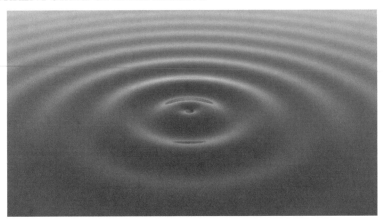

Figure 4.1: Expanding ripples in a pond.

occurring late in the process resulting in delays and added costs. Until recently the environmental impacts of the design product were rarely considered nor was there any consideration of societal impact given.

4.3   EMERGING TRENDS IN ENGINEERING: GREEN AND BEYOND

Bob Dylan composed "The Times They Are A-Changin" in the revolutionary days of the early 1960s—a period that spawned the anti-war movement, the civil rights and women's rights movements, as well as the environmental movement. Although written decades ago, his song could equally well describe today's growing concern about global climate change. As greenhouse gases trap more energy from the sun, and sea surface temperatures increase, sea levels also rise. Changes in ocean temperatures and currents brought about by climate change are leading to alterations in climate patterns around the world. In response to these complex issues, engineering is undergoing significant changes—particularly with respect to concern for the environment.

In the National Society of Professional Engineers code of ethics (Code of Ethics, 2020 [21]), it is stated, "Engineers in the fulfillment of their duties shall hold paramount the safety, health and warfare of the public."

This statement evokes a flurry of questions for our profession so let us consider several them.

- What is meant by the word "public"? Who or what does it include? Is it just the client? The immediate users? Organizations?

- Does "public" refer to only the members of society that are directly linked to the technology at the time that technology is developed?

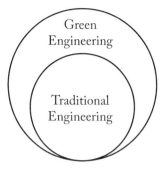

Figure 4.2: Traditional and green engineering.

- Or does engineering require a longer view in which the impact(s) of the technological advances on future generations are also considered?

- Does engineering have a responsibility to insure that people in the regions of the world marked by poverty have access to advances in technology?

- And finally, does "public" refer only to people? Should the engineer have any concern for the millions of species who inhabit the rest of our planet?

Different branches within the engineering profession have evolved because of seeking to answer these questions. These branches include the following.

1. **Green engineering** which seeks to deal with the challenges that have arisen due to the impact of technology on the environment.

2. **Humanitarian engineering** which has brought to the forefront the widening wealth gaps among the various parts of the world and the responsibility engineering has for addressing that growing split.

3. **Engineering combined with social justice** which focuses on the implicit political nature of engineering and engineering decisions.

4. **Omnium engineering**, a relatively new approach to engineering, which asks engineering to widen its concern to all species, not simply the human species.

Each new branch of engineering seeks to expand our sense of responsibility as professional engineers.

4.4   GREEN ENGINEERING

Overview

Green engineering seeks to design projects in a manner that reduces their environmental impact during design and fabrication activities as well as during the life cycle of the project (Fig. 4.1). Because it considers the health and well-being of the environment, green engineering extends the boundaries of the engineer's professional responsibility beyond those of the traditional approach.

History

The concept of green engineering grew out of environmental engineering which Buescher described as "one of the world's oldest professions" (Buescher Jr. [15]). Buescher detailed evidence of the practice of environmental engineering from the early Egyptian dynasties to the classical Greek and Roman civilizations. For instance, here are two green actions taken by the ancient Romans thousands of years ago (Barry, 2012 [8]).

- **The ancient Romans treated water and air as shared resources.**

  Plutarch wrote extensively about environmental issues and was quoted as saying "Water is the principle, or the element, of things. All things are water." Romans built aqueducts that carried clean water hundreds of miles to major population centers where it was distributed to the homes and businesses of those who could afford it. The legal code of the Roman emperor Justinian declared that, "By the law of nature these things are common to mankind—the air, running water, the sea and consequently the shores of the sea."

- **The ancient Romans used passive solar technology.**

  It was expensive to heat a home in ancient Rome because wood is a bulky fuel that was not readily available in much of the Roman Empire. Instead, the Romans burned coal, which was not only expensive but also dirty. Although the ancient Greeks first developed passive solar concepts, the Romans used their advanced engineering and design skills to improve the technique. Passive-solar buildings were (and still are) built based on the orientation of the sun's path, using the sun's rays to heat the interiors. The Romans even used glass to boost the solar gain of their buildings, capturing and storing the heat with masonry inside their homes, bathhouses, and businesses.

It was after World War II that environmental concerns became increasingly important in the United States. With the emergence of synthetic chemicals, many of the nation's waterways began to be covered in foam because many of these new chemical agents were not biologically biodegradable and existing water treatment plants were unable to breakdown this kind of waste. This period also witnessed an increased use of harmful pesticides. In 1962, Rachel Carson (Carson, 2008 [17]) published *Silent Spring* which alerted a large audience to the environmental and human dangers of indiscriminate use of pesticides, spurring revolutionary changes in the laws affecting air, land, and water. This book was a prime mover in the development of environmental awareness in the United States.

Within a decade of Carson's book, Earth Day was founded. Celebrated annually around the world on April 22, Earth Day is an event that demonstrates support for environmental protection. First celebrated in 1970, it now includes events coordinated globally by the Earth Day Network in more than 193 countries.

It is difficult for many students to imagine that before 1970, it was perfectly legal for a factory to spew black clouds of toxic smoke into the air or dump tons of toxic waste into a nearby stream. As environmental degradation increased in the 1960s and dense visible smog blackened the skies in many of the nation's cities and industrial centers, it became apparent that major environmental legislation was needed to address these issues. In fact, most of our current environmental statues were passed in in the period between the late 1960s and early 1980s.

- On January 1, 1970, **National Environmental Policy Act** (or NEPA), was passed, making the 1970s the "environmental decade." Legislation during this period concerned primarily first-generation pollutants in the air, surface water, groundwater, and solid waste disposal.

- Later in that same year the **Environmental Protection Agency** (EPA) was created, consolidating environmental programs from other agencies into a single entity.

- As issues concerning acid rain, visibility, and air quality were identified, the U.S. Congress passed the **Clean Air Act of 1970**, placing air pollutants such as particulates, sulfur dioxide, nitrogen dioxide, carbon monoxide, and ozone under regulation.

- The **Clean Air Act** was revised and expanded in **1977** and **1990**. The subsequent revisions improved the law's effectiveness and targeted newly recognized air pollution problems such as acid rain and damage to the stratospheric ozone layer.

- The **Endangered Species Act of 1973,** the primary law in the United States for protecting imperiled species, was designed to protect critically imperiled species from extinction as a "consequence of economic growth and development untempered by adequate concern and conservation."

Methodology

Green design teams commonly use an "eco-efficient" design methodology which can be summed up as being "less bad." This methodology seeks to be "less bad" by reducing toxic wastes, reducing energy consumption, and reducing the demand on various natural resources. It is a methodology that is subject to rules, regulations, and limitations.

Principles

While numerous authors have identified core principles of green engineering, one set of principles which is encompassing has been offered by Anastas and Zimmerman (Anastas, 2013 [3]). Their principles of green engineering are as follows.

- Ensure that all material and energy inputs and outputs are as inherently non-hazardous as possible.

- Strive to prevent waste rather than simply treating it or cleaning up after it is formed.

- Develop separation and purification operations that minimize energy consumption and materials use.

- Ensure that products, processes, and systems are designed to maximize mass, energy, space, and time.

- Reduce the amount of resources consumed to transform inputs into the desired outputs.

- Strive to minimize complexity.

- Target durability, not immortality.

- Strive to minimize excess.

- Minimize the diversity of materials to promote disassembly and value retention.

- Integrate material and energy flows using existing assets or energy sources with an emphasis on developing products, processes, and systems that require local materials and energy resources.

- Design products, processes, and systems that can be reused in their "afterlife" when they are no longer used as originally intended

- Choose materials and energy from renewable sources rather than finite reserves.

Current Status of Green Engineering

Green engineering is growing in acceptance throughout the engineering profession. The National Society of Professional Engineers' Code of Ethics now encourages engineers to adhere to the principles of sustainable development. The Code defines "sustainable development" as the "challenge of meeting human needs for natural resources, industrial products, energy, food, transportation, shelter, and effective waste management while conserving and protecting environmental quality and the natural resource base essential for future development."

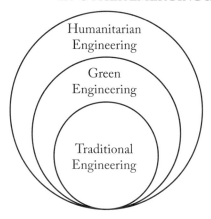

Figure 4.3: Traditional, green, and humanitarian engineering.

4.5   OTHER EMERGING IDEAS IN ENGINEERING

Humanitarian Engineering

Humanitarian engineering is "the application of engineering to improving the well-being of marginalized people and disadvantaged communities, usually in the developing world" with an emphasis upon sustainability, low costs, and locally available resources for the proposed solutions (Mitcham, 2010 [48]), as shown in Fig. 4.3.

Humanitarian engineering offers a view which is also more expansive than the traditional view of engineering. It states that it is not enough to consider the notion of "public" in a limited and narrow sense but rather encourages an integration of compassion, empathy, and trust toward our fellow human citizens. With its emphasis on developing communities and awareness of the cultural and societal differences that exist, humanitarian engineering's focus remains firmly affixed to human dignity, human rights, and fulfillment.

Engineering, Social Justice, and Peace

Engineering, Social Justice, and Peace (ESJP) is an emerging organization in engineering committed to envisioning and practicing engineering in ways that extend social justice and peace in the world (Engineering, Social Justice, and Peace, 2020 [33]); see Fig. 4.4. This commitment manifests in four major areas.

- Identifying and dismantling specific occurrences of injustice related to engineering and technology such as the use of child labor and sweatshops in the production of commercial goods.

- Devising and developing technologies and other engineering solutions broadly conceived to the problems they face in collaboration with community groups facing injustice.

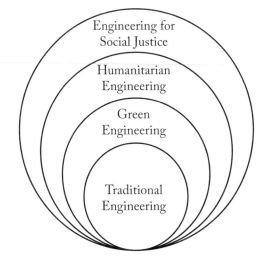

Figure 4.4: Traditional, green, humanitarian, and social justice engineering.

- Resisting injustice in its many forms through promotion of diversity and inclusivity, and by working toward fair, equitable, and sustainable treatment of people and their environments.

- Opposing globalized economic policies that lead to the destruction of local networks of labor, production, and food provision.

Omnium Engineering

In 2019, I proposed a new model or paradigm for engineering termed "omnium engineering" that considers the needs and wants of all life forms—not only those of the human species (Catalano, G., 2019 [18]). A literal translation of the Latin word *omnium* is "all" or "all beings."

Omnium engineering expands the boundaries of traditional engineering to include all of life (Fig. 4.5). It answers the question about who is in the "public" by asserting that all species must be included, not just the human species.

This new approach is based on advances in quantum mechanics, eco-philosophy, and complex systems.

- From quantum mechanics comes a view of the universe as a rich tapestry of potentialities existing in one connected system. This is in direct opposition to Newtonian mechanics which describes the universe as a mechanical and deterministic system composed of a collection of disparate objects (gears and levers).

- From eco-philosophy comes the idea of an evolving, dynamic universe governed by the principles of differentiation, communion, and subjectivity.

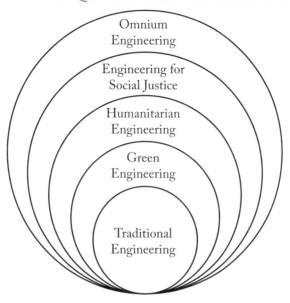

Figure 4.5: Traditional, green, humanitarian, social justice, and omnium engineering.

- From the study of complex systems come the concepts of nonlinearity, self-organization, and emergence. Nonlinearity means non-sequential or random. Emergence suggests the appearance of behavior that could not be anticipated from knowledge of the parts of the system alone. Self-organization means that there is no external controller or designer "engineering" the appearance of these emergent features; the features appear spontaneously.

4.6   QUESTIONS FOR THOUGHTFUL CONTEMPLATION

1. Compare and contrast green engineering with traditional engineering. How are they similar? How are they different?

2. Compare and contrast humanitarian engineering with engineering and social justice and with omnium engineering. How are they similar? How are they different?

CHAPTER 5

Team Dynamics: On Becoming a Cohesive Unit

"Your beliefs become your thoughts, your thoughts become your words, your words become your actions, your actions become your habits, your habits become your values, your values become your destiny."

Gandhi (Pinto, 1998 [55])

5.1 IMPORTANCE OF TEAMWORK IN ENGINEERING DESIGN

"A team is a group of people organized to work together interdependently and cooperatively to meet the needs of the client through the accomplishment of their purpose and goals."

(www.balancecareers/com)

Notice the emphasis on:

- organization,
- interdependence, and
- cooperation.

In 21st-century engineering, design is the task of teams—people working together toward a mutual goal, sharing ideas and experiences, and supporting each other throughout the process. An engineering design team must learn to collaborate; there is no room for competition or egos. For a team to work together successfully in a synergistic manner, a team member must become aware of her/his own individual strengths and weaknesses, personality traits, and values system.

Our Back Page: Teamwork is a Hiring Priority

The skills you will develop in capstone design will serve you well as you progress in your engineering careers. A recent listing of skills that engineering firms identify as most important includes:

- open-minded team player (at the top of the list!);

- strong technical knowledge;

- strong communication skills;

- low ego and high self-esteem; and

- ruthless prioritization and pragmatism (Ali, 2015 [2]).

5.2   STAGES OF TEAM DEVELOPMENT

As students begin their capstone design journey, they become part of a team with a shared mission—to solve a problem that has not been solved before. It is with one's team members that one will travel the yet unexplored terrain of engineering design for the first time. It is helpful to know some of the stages though which teams progress as team members learn to work together on a common goal.

The following model (FNSP) describes four stages of team development. It was developed in 1965 by Bruce Tuckman (Egolf, 2001 [32]).

Stage 1. Forming

In this stage, as individuals get acquainted, uncertainty and anxiety are often high. Individuals often wonder the following.

- "Is this a good fit for me?"

- "What does this team offer me?"

- "What do I have to offer this team?"

This stage generally is low in conflict and tensions because, as members begin to form impressions of each other, each wants to be accepted. This is a time of getting oriented—learning what to do, what is expected, what is acceptable, and how the group will operate.

Stage 2. Storming

This is frequently the most difficult stage of team interaction because it is generally a time of power struggle. As individual personalities begin to emerge, conflict and/or competition tend to arise. Dominating group members tend to emerge while the less confrontational members remain in their comfort zone. Questions often arise about leadership, authority, group structure, rules, and responsibilities. Values may clash and tensions can arise as the new team learns to work together.

The goal of this stage is not only to acknowledge and accept individual differences but also to identify one another's strengths and become cognizant of what each individual brings to the group. Only by working through these initial conflicts can the team begin to merge into a cohesive whole, a true team working toward a single purpose.

Stage 3. Norming

In this stage, as consensus develops about the leader of the team, a sense of cohesion and unity emerge. The interaction of the group becomes more enjoyable as cooperation rather than competition is accepted as the guiding principle. There is more open communication, bonding, and mutual respect, and if conflict arises, it is relatively easy to resolve in order for the group to move forward.

It is at this point that the group tends to develop "team norms"—a set of standards for behavior, attitude, and performance that all of the members are expected to follow. Although norms are not written down, they are implicitly understood by the team.

Stage 4. Performing

In this stage, consensus and cooperation among the team members are well-established, and the group morale is high. The team is organized, mature, and well-functioning. Group members actively acknowledge the strengths, skills, and experience that each member brings to the group. The individual members are flexible and willing to adapt according to the needs of the group. There is a clear structure in the group, a sense of trust and interdependence, and total commitment to the success of the team's mission.

5.3   PREPARING TO WEATHER THE STORM

In order to survive the "storming" phase of team development, it is helpful for each individual team member to gain insight into his or her own personality and values in order to become aware of them in others. There is a variety of ways available to examine individual personal traits and develop an understanding of how those traits or qualities may affect team dynamics. Two of the more common methods include the following.

1. Myers–Briggs Personality Inventory (MBTI)

   The MBTI is an introspective self-report questionnaire that identifies and clarifies how different people perceive the world around them and make decisions (Quenck, 2009 [56]). It is based on the conceptual theory proposed by Carl Jung who speculated that humans experience the world using four principal psychological functions—sensation, intuition, feeling, and thinking—and that one of these four functions is dominant for a person most of the time. The MBTI offers an opportunity both for self-reflection and for opening a dialogue with other design team members.

2. Five Factors (OCEAN) Model (Wiggins, 1996 [70])

The five-factor model of personality (FFM) is a set of five broad trait dimensions or domains, often referred to as the "Big Five": Openness, Conscientiousness, Extraversion, Agreeableness, and Neuroticism. The FFM was developed to represent as much of the variability in individuals' personalities as possible, using only a small set of traits. By identifying the factors—traits—which best describe them as individuals, team members gain a better understanding of themselves as well as of each other.

The five factors more fully described here.

- O: "Openness to experience (inventive/curious vs. consistent/cautious). Appreciation for art, emotion, adventure, unusual ideas, curiosity, and variety of experience. Openness reflects the degree of intellectual curiosity, creativity and a preference for novelty and variety a person has."

- C: "Conscientiousness (efficient/organized vs. easy-going/careless). Tendency to be organized and dependable, show self-discipline, act dutifully, aim for achievement, and prefer planned rather than spontaneous behavior."

- E: "Extraversion (outgoing/energetic vs. solitary/reserved). Energetic, urgency, assertiveness, sociability and the tendency to seek stimulation in the company of others, and talkativeness."

- A: "Agreeableness (friendly/compassionate vs. challenging/detached). Tendency to be compassionate and cooperative rather than suspicious and antagonistic toward others. It is also a measure of one's trusting and helpful nature, and whether a person is generally well-tempered or not."

- N: "Neuroticism (sensitive/nervous vs. secure/confident). Tendency to be prone to psychological stress. The tendency to experience unpleasant emotions easily, such as anger, anxiety, depression, and vulnerability. Neuroticism also refers to the degree of emotional stability and impulse."

3. Other excellent resources available online include the following.

- *Mindtools.com:* useful compendium of online resources for project management and leadership development.

- *Leadership Coaching and Training.com:* training in leadership development and communication skills.

- *The People Equation.com:* training in leadership development as well as time and stress management techniques.

5.4   CORE VALUES

"Our values are the lens through which we view the world: they stem from our underlying beliefs and assumptions which are generally neither articulated nor questioned."

(Baillie, 2012 [7])

A value is a fundamental belief that guides and influences one's thoughts, actions, attitudes, and feelings. These basic beliefs result from a wide variety of factors including one's family, religion, peers, culture, race, gender, and social background. Some examples of values include dependability, reliability, loyalty, commitment, honesty, open-mindedness, efficiency, innovation, creativity, compassion, courage, perseverance, and service to others.

Because they determine what is important to us—what matters, they serve as the basis for moral codes and ethical reflection not only for individuals but also for societies and cultures, communities, institutions, and professions. Unfortunately, many people fail to examine these basic beliefs, and instead, simply take them for granted.

Because engineers are asked to make scientific and engineering judgments in every step of the design process, Mitchell and Baillie point to the need for engineering students and practicing engineers to identify their own values and perspectives in order to recognize what influences their point of view when problem-solving. As each team member begins to understand him/herself, a greater understanding and tolerance of others inevitably develops, and the foundation for an effective team is forged.

5.5   VALUES CLARIFICATION EXERCISES

Although there are many different activities that have been developed to help clarify one's values, we have selected two to share with you. Each forces a careful examination of what matter to each of us as individuals.

- **The List-Outcome-Behavior Exercise**

This particularly useful approach was created by the University of Wisconsin (Core Values Clarification Exercise, 2016 [23]).

1. Using eight words from an extensive list of values, each individual selects, followed by a narrowing to three to five. A partial list of the values may include:

   - wisdom;
   - integrity;
   - trust;
   - beauty;

- faith;

- respect;

- love;

- honesty;

- caring; and

- service.

2. Each individual is then asked to develop an outcome statement and subsequent behavior for each value. For example, if "service" was identified as a high priority, the individual's outcome statement might be "become involved with different volunteering activities or non-profit organizations." The resultant behavior then might be "put personal interests aside to give more freely to others."

3. Each individual then processes the results by responding to questions such as:

- Is there consistency between what was identified as core values and personal behavior?

- Are there gaps between the two? What might explain those gaps?

- **The Kidney Transplant Exercise**

In the following exercise it is important that you respond first as an individual and afterward discuss your responses with the rest of your team. That discussion will add to your understanding of both yourself and your teammates.

You find yourself at a board of directors meeting at a local hospital. The meeting begins: "Tonight, the Laymen's Board of Review of General Hospital" meets to consider applicants for kidney transplants. Each of the patients described below has been evaluated by the medical staff, and it has been determined that each patient will probably die in three to six weeks without a transplant. The best statistical estimates are that only about five donors will be available during that period. The board must establish a priority list of who will get the kidneys by rank-ordering the nine applicants. They are:

1. John Hallbright. Age 41. Married. Two children, a son 12 and a daughter 4. College graduate. Works as an officer in a bank. Wife also employed as an elementary school teacher.

2. Marie Villareal. Age 39. Unmarried. College graduate; holds a Master's in physical therapy. Employed at VA hospital 14 years; is head of Physical and Occupational Therapy treatment center.

3. Pamela Watson. Age 23. Married, no children. College graduate. Teaches social studies and is cheerleader advisor in junior high school. Husband is a high school teacher. Medical diagnosis indicates a heart condition that may cause complications in a transplant operation.

4. Avery Smith. Age 51. Married. Three children, a daughter 19 and two sons 17 and 15. High school graduate. Owner and operator of Smith Industries, a machine shop that employs 150. City councilman for 12 years; member of library board of directors for 6 years.

5. William Work. Age 11. One of seven children of Mr. and Mrs. Ralph Work. Has received a kidney transplant that failed.

6. Walker Red Cloud. Age 22. Ojibwa Indian. At least four children by two wives. Fourth grade education. No occupation.

7. Nancy Adams. Age 34. Divorced. Three children, a daughter 7 and twin sons, all in her custody. Employed as a secretary in a real estate office. Receives no child support from her ex-husband, whose whereabouts are unknown.

8. Mary Parenti. Age 12. IQ 87. Teachers describe her as shy, withdrawn, and inhibited. Family emigrated to New York the year she was born. Family owns restaurant where both parents work.

9. Juan Gonzalez. Age 32. Married, eight children. Migrant worker. Wife and three oldest children also work as migrant workers.

Only five individuals can receive kidney transplants. Pause a moment and make your decision as a board member. Which five would you select?

Now as an individual:

- How did you make your decisions?

- Did you establish criteria in advance for making decisions?

- Can you specify what the criteria are?

- Did you make assumptions that were based on stereotyping? Generalizations?

Now as a team:

- Can you reach a consensus? What did you learn from the experience of trying to reach a consensus?

Figure 5.1: Teamwork (Bobel, 2018 [12]).

Our Back Page: Lessons from the Pack

The photograph in Fig. 5.1 has a very special significance for us in the teaching of capstone design. The photo depicts a wolf pack traveling through the winter snows of Yellowstone National Park. What is most significant is the order of the wolves in the pack. The alpha male and female wolves are at the back of the pack making certain that all are safe with no one left behind. The old and the sick wolves are in the front of the pack setting the pace followed by with the younger wolves in the middle.

Because cohesion, teamwork, and training of the pack determine whether the pack lives or dies, wolves have mastered the technique of focusing their energies toward the activities that will lead to the accomplishment of their goals. They have a strategic plan and execute it through constant communication.

Not everyone in the pack strives to be the leader in the wolf pack. While some are consummate hunters, others act as caregivers or even jokesters, with each gravitating to the role that emerges from playtime as a pup. A wolf's behavior is always based upon the question, "What is best for the pack?"

WE THINK THE PARALLELS ARE OBVIOUS!

5.6   QUESTIONS FOR THOUGHTFUL CONTEMPLATION

1. Take an online version of the **MBTI** or the **Five Factors** questionnaires. (There are quite a few available for free!)

   (a) After you receive your results, write about your results and how they both surprised you and confirmed what you already knew about yourself.

   (b) Share what you are comfortable in sharing from your essay with your teammates.

   (c) What insights concerning the members of your team result from the sharing? Has the chemistry of the team changed in any way? How so?

2. Complete the **List-Outcome-Behavior** exercise. List at least ten values that are important to you. What insights does the exercise provide for you?

3. Answer the questions posed at the end of the **Kidney Transplant** exercise. After reflecting on your responses, what insights does the exercise provide for you?

CHAPTER 6

Design Methodology

"It's not 'us vs. them' or even 'us on behalf of them.' For a design thinker it has to be 'us with them"

Tim Brown, CEO and President of IDEO

6.1   WHAT IS DESIGN METHODOLOGY?

While scientists ask questions and develop experiments to test those questions, engineers use the engineering design process to create solutions to problems. The design process utilizes a systemic approach—called a method—to solve an open-ended problem.

Engineers are not limited to one methodology; in fact, there are quite a few available including the following.

- **Traditional linear**: focuses on the end-product.

- **Concurrent**: utilizes many of the same steps as the traditional approach, but requires that all steps be in progress at the same time (i.e., concurrently).

- **Eco-efficient**: focuses on doing less damage to the environment.

- **Eco-effective**: focuses on not only doing less damage but improving the health of the environment.

The methodology selected will vary from project to project and team to team. Not one size fits all! The final selection requires that a consensus be reached by all individuals who are involved—team members and client(s) alike.

6.2   THE TRADITIONAL LINEAR METHOD

The traditional engineering design process typically focuses upon the end-product. This methodology generally utilizes the following five steps (Fig. 6.1).

1. *Define the problem*

   This involves meeting with client(s) and clearly identifying and defining the problem and includes listing the product and/or customer requirements as well as specific information about product functions and features.

Figure 6.1: Flow chart for traditional engineering design.

2. *Gather pertinent information*

   Using a variety of resources, the team obtains relevant information for the design of the product and its functional specifications.

3. *Generate multiple solutions*

   A variety of strategies and tools are utilized to achieve the goals and the requirements of the design.

4. *Analyze and select a solution*

   A detailed analysis is created which allows for a thorough study of each of the possible solutions—resulting in the identification of the final design that best fits the product requirements.

5. *Test and implement the solution*

   A prototype of the design is often constructed, and functional tests are performed to verify and possibly modify the original design.

   Because unexpected issues and/or unanticipated results often occur during the testing stage, design steps are frequently repeated. The first solution may prove unworkable for any number of reasons and may require redefining the problem, collecting more information, and/or generating different solutions. Design can be a cyclical process in this respect.

6.3 CONCURRENT ENGINEERING DESIGN

The basic premise for concurrent engineering (Fig. 6.2) revolves around two concepts:

1. all elements of a product's life-cycle—from functionality, production, assembly, testing, maintenance, environmental impact, and finally disposal and recycling—need to be taken into careful consideration in the early design phases; and

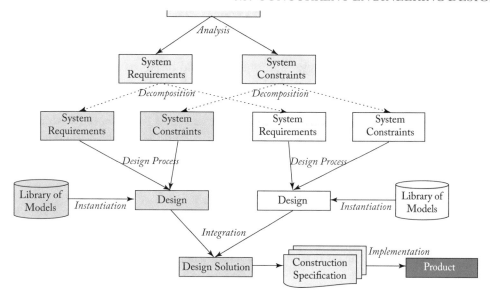

Figure 6.2: Concurrent engineering design approach (Tudorache, 2005 [65]).

2. all the design activities need to be occurring at the same time—i.e., concurrently.

The advantages of concurrent engineering over a traditional engineering approach include the following.

1. It encourages multi-disciplinary collaboration.

2. Product cycle time is reduced.

3. Cost is reduced.

4. Quality is enhanced and supported throughout the entire project cycle.

5. Productivity is increased because issues and mistakes are identified and overcome earlier.

The disadvantages of concurrent engineering are as follows.

1. It is complex to manage.

2. Teamwork and open, effective communication are critical.

3. A mistake or error in one aspect of design is directly linked with all other aspects of the design.

6.4   ECO-EFFICIENT ENGINEERING DESIGN APPROACH

Over the years, as countries and regions around the world began to develop factories and industries, it slowly became evident that the health of the Earth's eco-systems was being negatively affected by increasing technology and that engineering had to respond to that threat. The methodology of eco-efficiency evolved as one of the main tools to promote the shift from unsustainable development to sustainable solutions. This method is based on the dual objectives of

1. creating more goods and services while using fewer resources and

2. creating less waste and pollution.

An eco-efficient engineering design method is one that is consistent with the principles of green engineering. This means it addresses the following green engineering principles.

- Reduce wastewater.

- Reduce energy usage.

- Reduce the use of natural resources.

- Meet all rules, regulations, and limitations aimed at reducing or slowing down environmental damage.

- Strive to be "less bad" by doing less damage to the Earth than traditional approaches.

6.5   ECO-EFFECTIVE ENGINEERING DESIGN APPROACH

McDonough and Braungart challenged the eco-efficient notion of minimizing the impact on the environment by asking the provocative question: When is being "less bad" not good enough?

They propose a design methodology described as "cradle-to-cradle" that is based on eco-effectiveness, rather than eco-efficiency (i.e., being "less bad") (Braungart, 2002 [14]). Their framework seeks to create production techniques that are not just efficient but are essentially waste free.

In "cradle-to-cradle" production, after products have reached the end of their useful life, they become either technical nutrients that can be recycled or reused, or they become biological nutrients that can reenter the environment by being composted or consumed. This has been called "upcycling" or "creative reuse" because it aims to transform by-products, waste materials, and useless or unwanted products into new materials or products of better quality.

By contrast, "cradle-to-grave" refers to the "downcycling" methodology used by the eco-efficient model of "reduce, reuse, recycle." Downcycling is the recycling of waste where the recycled material is of lower quality and functionality than the original material. Here a company may take responsibility for the disposal of goods it has produced, but either does not place the product's constituent components back into service or places a lesser product back into use.

The principles of the "cradle-to-cradle" design include:

- recognizing in nature the discharge of one system becomes food for another (i.e., waste=food);

- utilizing renewable energy such as solar, wind, geothermal, and gravitational energy; and

- celebrating diversity of the natural world and of the myriad varieties of cultures.

To illustrate a process or industry which utilizes an eco-effective model of design, McDonough and Braumgart describe a community of ants whose daily activities include:

- handling their own material wastes and those of other species in a safe and effective manner;

- growing and harvesting their own food while nurturing the ecosystem of which they are a part;

- constructing houses, farms, dumps, and food-storage facilities from materials that can be recycled;

- creating disinfectants and medicines that are healthy, safe, and biodegradable; and

- maintaining soil and health for the entire planet.

Our Back Pages: Cradle-to-Cradle

One industrialist who championed this approach to manufacturing was Ray Anderson. He had a radical vision that extended far beyond complying with environmental standards or placating investors and customers with a push to "go green." Anderson, the founder and CEO of the Atlanta-based carpet manufacturer *Interface Inc.*, wanted his company to give back more than it took from the earth. Anderson often stated, "I challenge the company to go beyond sustainable—to put back more than we actually take from the earth, and to do good for the earth, not just harm."

6.6   QUESTIONS FOR THOUGHTFUL CONTEMPLATION

1. Compare and contrast a traditional engineering methodology to a concurrent engineering methodology. List and describe the similarities and differences.

Figure 6.3: Ray Anderson, A Hero for the Environment (Vitello, 2011 [68]).

2. Compare and contrast an eco-efficient engineering methodology to an eco-effective engineering methodology. List and describe the similarities and differences.

3. Research the contributions of Mr. Ray Anderson. Why did the *New York Times* refer to him as a "Hero for the Environment?" Identify someone you consider a "Hero for the Environment." Explain why you chose that person.

CHAPTER 7

Holistic Design Tools

> *"The ideal engineer is a composite … He is not a scientist, he is not a mathematician, he is not a sociologist or a writer; but he may use the knowledge and techniques of any or all of these disciplines in solving engineering problems."*

Nathan W. Dougherty (Mades, 2018 [47])

7.1   THE TOOLBOX

No matter which design methodology is jointly selected by your team and your client, you will need a wide range of design tools to support your project from inception to successful conclusion and delivery. In this chapter we are offering a set of holistic design tools—tools that are applicable to all design projects.

Another set of tools called DESIGN FOR EXCELLENCE can help address concerns for a specific aspect of the design's process and makeup. These are offered in Chapter 8.

Every design toolbox needs to include the following holistic tools.

- *Client statement:* This written statement is formulated after the initial meeting with the client and is a description of the design problem as stated by the client. A key ingredient in formulating this statement is the establishment of trust among the team as well as with the client(s).

- *Statement of Work (SoW):* This follows the client statement and details the deliverables of the project as well as goals, milestone schedules, costs estimates, etc.

- *TRIZ:* This tool transforms the statement of work into actual engineering *Design Objectives*. TRIZ enables you to identify the *Ideal Final Result (IFR)*—the best solution that can be imagined with all judgments suspended at the outset.

- *The Pairwise Comparison Chart:* This tool is used to sort through the relative importance of the design objectives so they can be ranked from most to least important.

- *A Function Means Analysis (FMA):* This tool transforms the prioritized design objectives from abstract objectives into actual possible means for accomplishing those objectives.

Figure 7.1: Modern designers toolbox.

- *Black and Transparent Box diagrams:* These diagrams enable the team to develop a more complete picture and a better understanding of all the tasks that need to be accomplished in meeting the design objectives for the project.

- *Rich Picture:* This tool provides a visual representation for the project, client, and environment.

- *CATWOE:* This tool is a simple checklist that can help you and your team visualize the "big picture" near the mid-point of the project to remain focused.

7.2   MEETING WITH YOUR CLIENT AND WRITING A CLIENT STATEMENT

After selecting a project, the team's first step on the design journey is the meeting with the client. The goals of this initial meeting include:

- identifying the client's needs and expectations;

- identifying the challenges both the team and the client face; and

- developing a relationship based on trust with your client.

Our Back Pages: What is trust?

To trust means to believe in the reliability, truth, ability, or strength of a given being or object. It means that we have confidence in the intentions and motives of the other party.

Stephen Covey wrote: "When the trust account is high, communication is easy, instant, and effective." Pointing out that trust is essential to healthy relationships, he also wrote, "Trust is the glue of life" (Covey, 2013 [24]).

Trust-building is an essential skill in today's world of uncertainties. Here are some strategies you will find helpful in establishing trust—not only with your client(s) but also with you team members and others whom you encounter in life.

- Be well prepared.

- Be respectful, courteous, approachable, and friendly.

- Show interested in the other party. Ask questions and then listen carefully.

- Listen some more.

- Be considerate of others and their feelings. Respect their ideas and perspectives.

- Most importantly, keep your word and follow through with your actions—not just some of the time, but ALL of the time.

After the initial meeting the Client Statement is formulated by the team based on the in-person interviews and exchange of several draft statements outlining the project goals. Keep in mind that many clients are not engineers. Part of the challenge in the writing of a strong Client Statement is recognizing that team members and clients may have a different set of life experiences, different ages, and even a different vocabulary.

This is an example of a client's original statement to the design team:

"Dynamic cyclic stretch and electrical stimulation have been independently shown to improve strength, maturation, and function of engineered cardiac tissues. However, most current bioreactors incorporate either dynamic stretch or electrical stimulation, but few integrate both stimuli simultaneously. Commercially available bioreactors are too expensive and are typically only compatible with one tissue geometry. We need a a low-cost bioreactor that can simultaneously impart electrical stimulation and cyclic stretch to ring-shaped engineered cardiac tissues."

The team's subsequent client statement is as follows:

"The client desires a novel bioreactor for growing and supporting fetal cardiomyocyte cells under simultaneous mechanical and electrical stimuli. Electrical impulses should have voltages and frequencies which can be varied based on the needs of researchers, with ranges including physiologically relevant values. Mechanical strain magnitude and frequency should also be variable and include a range of physiologically relevant values. Cells grown in the bioreactor must have adequate exposure to the medium being used, as well as dissolved gasses. There will be a base budget of $800 for this project, though additional support will be available from the lab."

The final Client Statement is no longer general but much more specific. The actual budget is specified as well as the physical parameters and their proposed variations. Although this process of taking a client's initial statement and adding as much technical information as possible takes time and effort, a strong Client Statement that will help in the development of a strong Statement of Work which comes next.

7.3   STATEMENT OF WORK

After meeting with your client and developing a strong Client Statement, the next step is to write a Statement of Work (SoW). This is a narrative description of a project's work requirement that defines project-specific activities, deliverables, and timelines.

There are multiple versions of SoWs available online. As a minimum, the SoW document should clearly detail the following.

1. What is the project? Why it is necessary? What is its purpose? What it will achieve?

2. Who has approval (governance)?

3. How will the project be completed? What methodology/approach do you plan to use? What tasks are involved?

4. What will be produced? What are the deliverables?

5. When will the project be delivered? What is your timeline? What are your milestones?

6. What is your estimated cost? What is the payment schedule?

7. What assumptions are made in the design process?

7.4   TRIZ AND THE IDEAL FINAL RESULT (IFR)

TRIZ is the Russian acronym for the "Theory of Inventive Problem Solving" which was developed in 1946 (Mulder, 2016 [49]). It is a creative problem-solving tool used in product and process development that culminates with a solution called the **Ideal Final Result (IFR)**. An IFR is the best solution that can be imagined when all judgments have been suspended.

Here are three examples of TRIZ diagrams from student design teams (see Figs. 7.2–7.4). A generalized step-by-step approach in using TRIZ is shown in Fig. 7.5.

- Step 1: The problem statement is stated in the center of a conceptual map.

- Step 2: Elements of an Ideal Final Result (IFR) are located around the circumference of the problem statement. These elements will later be identified as the **Design Objectives**.

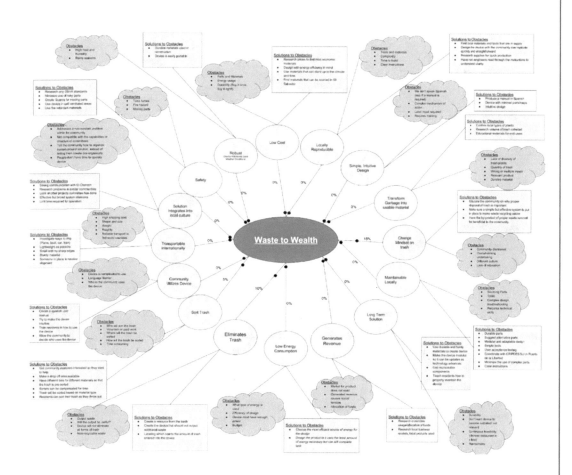

Figure 7.2: TRIZ diagram for waste-to-wealth project.

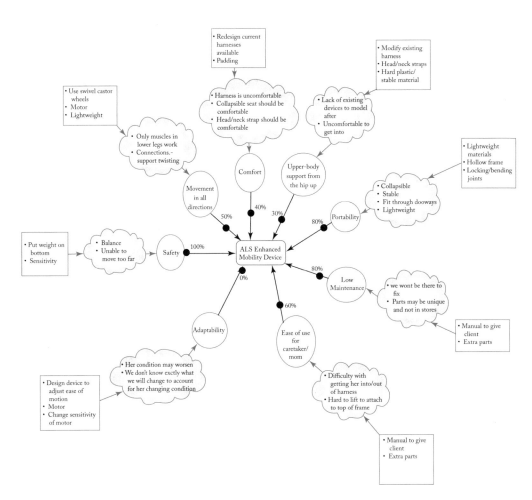

Figure 7.3: TRIZ diagram for ALS enhanced mobility project.

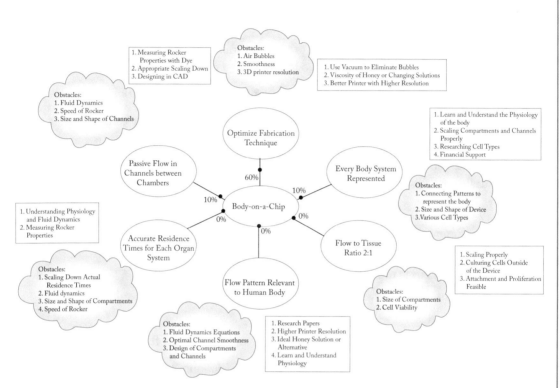

Figure 7.4: **TRIZ** diagram for body-on-a-chip project.

Figure 7.5: Four steps in construction of a TRIZ diagram.

- Step 3: An assessment is made as to how close or how far away present technology is for each of the elements identified in Step 2. Note that 0%-located at the problem statement—would indicate that at present that element of an Ideal Final Result is completely lacking while 100%—located at the location of the element—would indicate the present technology incorporates that element completely.

- Step 4: Obstacles preventing the present technology from achieving a 100% solution would then be identified would be indicated in a cloud near each element.

The following example illustrates how to use this tool in a design problem.

- Step 1: Your team is presented with problem of transforms our campus into a completely sustainable ecosystem.

- Step 2: First your team would need to identify the elements of such a campus—the Ideal Final Result (IFR).

Let us limit this example to five elements.

   – Renewable energy used for all electrical power generation as well as heating and air conditioning.
   – Local farms used as sources for all food stuffs.
   – Unused food served in eating facilities given to local food banks.
   – Any/all solid wastes recycled.
   – All student work and office work done online without the need for paper.

- Step 3: This step involves "estimating the gap"—judging how far away your campus is from each of these goals. This will require your most informed guess.

   – Suppose your campus already has a few solar cells but still utilizes coal fired plants. Maybe you would only be about 25% to the goal of completely renewable energy sources.
   – Since most of your food comes from local farms in part of the academic year, in the dead of winter you rely on foodstuffs from around the globe. Maybe you are at about 50% of our goal.
   – Then there is the excess food question. Due to health and insurance regulations, suppose your campus is forbidden from transporting food off campus. You are at 0% with this element.
   – Recycling is another challenge. Your campus might have a recycling program, but it suffers from lack of visibility. So maybe this one hovers around 10%.
   – Last, there is the paper free requirement. This one seems to be going well and you are now about 90% there.

Compiling all this information into one diagram—the IFR—allows you to imagine an ideal solution to your design problem and to consider your current standing. At this point you are ready to list ways to overcome any obstacles.

The end-product of your TRIZ implementation is:

- a visual representation of the Ideal Final Result for the design problem that incorporates elements of what would be an ideal or ultimate result,

- an assessment of where presently available technology matches the IFR, and

- an accounting of possible obstacles.

You can then use the described elements are then used as the basis for **Design Objectives** used in the next steps in a design process.

7.5 DESIGN OBJECTIVES AND THE PAIRWISE COMPARISON CHART

The TRIZ construction of the IFR allowed you to identify the important elements of your proposed design. These elements are called **Design Objectives**. In the example of the sustainable campus, we included five elements of equal importance. Actual design problems may have many, many more design objectives, not all equally important. The **Pairwise Comparison Chart**, described by Dym and Little (Dym 2008 [31]) allows the design team to prioritize the different objectives from least to most important.

To construct a Pairwise Comparison Chart, a matrix is created with the design objectives listed horizontally in the first row and vertically in the first column. Each column objective is then compared to each row objective with a rubric for comparing one to the other. Typically, this rubric involves using +1 in the matrix location if the column objective is more important than the row objective, 0 if neither is more important, and -1 if the column objective is less important. The matrix objectives are then added across with the higher numbers indicating more importance. By examining the entire matrix then the designer can begin to prioritize the design objectives.

Below is an example constructed by the **Waste-to-Wealth** team (Fig. 7.6). Note the range of numbers in the total column range from +15 for safety to -13 for generating profits in this case. It quickly becomes clear which of the design objectives are the most important and which has less importance.

One approach to generating the final version of this pair-wise comparison of different design objectives is for the design team as well as the client(s) to construct separate charts and then meet to discuss and agree to the final rankings. Such a discussion can serve to ensure the team's approach is consistent with the client(s)'s goals for the project. The **Waste-to-Wealth** team's pair-wise comparison chart shown above is an example of such a discussion.

Design Objectives	Intuitive Design	Maintainable locally	Low cost	Locally Reproducible	Change mindset on trash	Transform garbage into usable material	Long term solution	Generate revenue/profit	Low energy use	Eliminate trash	Sort Trash	Community utilizes device	Transportable	Integrates with Culture	Robust	Safety	Total
Intuitive Design	■	-1	-1	-1	1	-1	-1	-1	-1	-1	1	-1	1	0	-1	-1	-8
Maintainable locally	1	■	1	0	1	-1	0	1	0	-1	1	-1	1	1	-1	-1	2
Low cost	1	-1	■	-1	1	-1	-1	1	-1	-1	1	-1	1	1	-1	-1	-3
Locally Reproducible	1	0	1	■	1	0	1	1	0	-1	1	-1	1	1	-1	-1	4
Change mindset on trash	-1	-1	-1	-1	■	-1	0	1	-1	-1	1	0	1	0	-1	-1	-6
Transform garbage into usable material	1	1	1	0	1	■	1	1	1	-1	1	0	1	1	1	-1	9
Long term solution	1	0	1	-1	0	-1	■	1	-1	-1	1	-1	1	-1	0	-1	-2
Generates revenue/profit	1	-1	-1	-1	-1	-1	-1	■	-1	-1	-1	-1	-1	-1	-1	-1	-13
Low energy use	1	0	1	0	1	-1	1	1	■	-1	1	-1	1	1		-1	4
Eliminate trash	1	1	1	1	1	1	1	1	1	■	1	1	1	1	1	-1	13
Trash is sorted	-1	-1	-1	-1	-1	-1	-1	1	-1	-1	■	-1	1	-1	-1	-1	-11
Community utilizes device	1	1	1	1	0	0	1	1	1	-1	1	■	1	1	0	-1	8
Transportable internationally	-1	-1	-1	-1	-1	-1	-1	1	-1	-1	-1	-1	■	-1	-1	-1	-13
Solution integrates with local culture	0	-1	-1	-1	0	-1	1	1	-1	-1	1	-1	1	■	-1	-1	-5
Robust	1	1	1	1	1	-1	0	1	0	-1	1	0	1	1	■	-1	6
Safety	1	1	1	1	1	1	1	1	1	1	1	1	1	1	1	■	15

Figure 7.6: Pairwise comparison chart for waste-to-wealth design project.

Main Function: Transform Garbage into Usable Material												
Collect Garbage	Feed Into Machine	Grinder/ Shredder	Collect Output	Heating Element	Melted Plastic Holder	Apply Force	Create Mold	Collect Output	Cooling Mechanism	Produce Product	Ventilation	Insulate Heated Parts
Family group sorts trash	Conveyor belt	Circular metal wheel	**Bucket/ container**	Fire	Metal pan container	Hydraulic	**Metal**	**Collection bin**	**Air**	Produce fiber	Ducts	Spray foam
Committee sorts trash	Manually	Manual paper cutter	Conveyor belt	**Electric heating coil**	Spool	**Lever-based press**	Concrete	Conveyor belt	Fans	**Mold plastic**	Fans	**Rubber**
Trash is brought to a central location to be sorted	**Funnel**	**Interlocking gears**	Hands	Convection oven	**Right into mold**	Pneumatic air compressor	Wood		Ice		Vents	Fiberglass
Communal trash cans		Paper Shredder					Ceramic		Water		**Open area**	

Figure 7.7: Function means tree for waste-to-wealth design project.

7.6   FUNCTION MEANS TREE

Once you have identified and ranked the different design objectives, it is time to create a **Function Means Tree** (Dym, 2008 [31]). This is a tool that uses a tree diagram to breakdown broad categories of tasks into small sub-tasks. For each sub-task, a range of ways to accomplish the sub-tasks is listed (see Fig. 7.7).

 To create a function-mean tree, first write your project goal at the top. Then, in the top row of the matrix, list all the functions or tasks in separate columns. Next, underneath each task or function, list all the different means to accomplish each task; the more options, the better. Finally, the design team can select a means for each of the different functions or subtasks and watch as a design solution emerges. In the final step the team selects and shades entries in each column to form a workable design solution.

7.7   RICH PICTURE AND CATWOE ANALYSIS

At this stage in the design journey, it is often useful to take a step back and take a much wider view of the entire process. The **Rich Picture** (Williams, 2020 [71]) and **CATWOE** Analysis (CATWOE Analysis [20]) are helpful tools used to gain a wider perspective of the current state of the project.

 Creating a Rich Picture is a way to explore, acknowledge, and define a situation and express it through diagrams to create a preliminary mental model. Developed as part of the **Soft Systems Methodology**, a Rich Picture is a drawing that illustrates the main elements of the design and the relationships that need to be considered. Not only does it help to open discussion that can lead to a broader shared understanding of a project, it can also be used to provides a "big picture" of the product and identify the range of individuals involved and the roles each plays.

Consider the application of the Rich Picture to the **Waste-to-Wealth** project (Fig. 7.8). The entire scope of the project is visually portrayed from the collection process through the sorting stage and eventually into the transformation of the waste to viable, usable products. The diagram illustrates significant roadblocks along the way as well as the steps in which people are involved with the design solution. It not only helps the team identify strengths and weaknesses of their proposed design, but it also keeps the client(s)' needs visible and in the foreground at all times.

Below is the application of the Rich Picture to the **ALS Enhanced Mobility Device** project (Fig. 7.9). Here also the "big picture" of the project is visually portrayed with the client's needs visible and in the foreground at all times. This Rich Picture also includes members of the client's community whose views and perspectives are to be valued and considered. The team's graphic points out significant roadblocks along the way as well as the steps involved with the design solution. It also helps the team identify strengths and weaknesses of their proposed design.

CATWOE is a listing of all the characters, relationships, and contexts visually described in the Rich Picture.

CATWOE is an acronym that breaks down as follows:

C: Clients, stakeholders

A: Actors who are actively making changes

T: Transformation: the outcome of the design

W: Weltanschauug/world view: the "big picture"

O: Owners

E: Environmental factors that can influence the outcome (i.e., ethics, regulations, finances, environmental issues.

Typically during the conceptual phase it is helpful to stop and utilize this analysis tool together with the Rich Picture to gather the perceptions of all the different individuals involved in the project (all stakeholders) in a common platform. The combination of CATWOE and Rich Picture provides a holistic way to incorporate their different perspectives in the environmental, societal, and political contexts in which the entire project is taking place.

Below is the application of CATWOE to the **Waste to Wealth** project. Below is the listing for each of the letter identifiers:

- Clients: Their clients are the villagers and local town officials.

- Actors: The actors are the villagers.

- Transformation: The plastic waste is transformed into a material that can be used in various fabrication and construction applications.

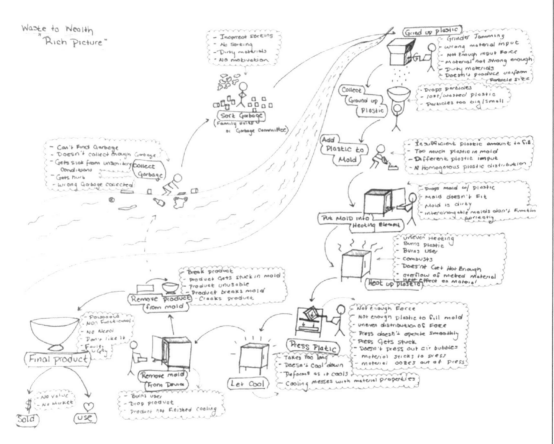

Figure 7.8: Rich picture for waste-to-wealth design project.

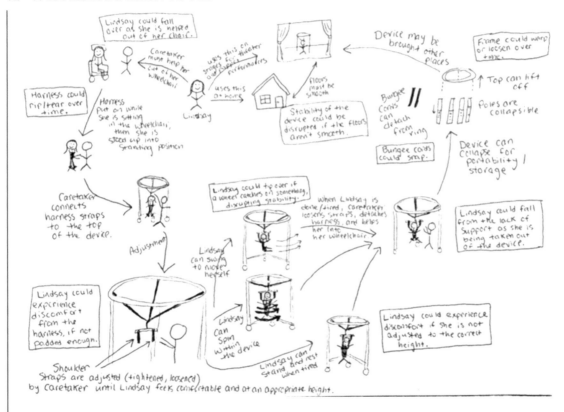

Figure 7.9: Rich picture for ALS enhanced mobility device.

- Owners: Villagers and local town officials are the owners of the product.

- Worldview: The village has struggled with pollution from discarded plastic bags and other materials for many years. It not only poses a health problem not only for the village itself but also those communities located downstream.

- Environment: Local, historical attitudes concerning recycling and the lack of suitable alternatives for the plastic discarded contribute to the problem.

Consider the application of CATWOE to the **ALS Enhanced Mobility** project.

- Clients: Artist and her supporting family and friends.

- Actor: Artist.

- Transformation: With the use of the device, the client is now able to paint using her feet as well as dance. The device has the potential to bring great joy back into the client's life.

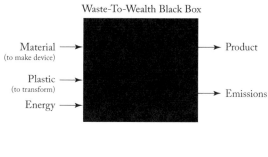

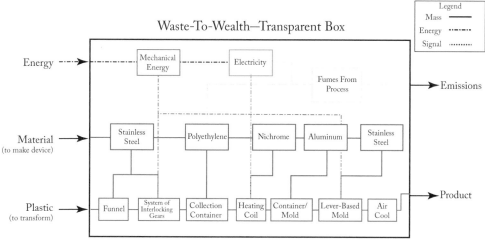

Figure 7.10: Waste-to-wealth black and transparent boxes.

- Owners: Artist and her family

- Worldview: The client has incredible strength and a passion for living life to its fullest no matter the obstacles. She inspires all with whom she comes in contact.

- Environment: The family is incredibly supportive as are her close friends and neighbors.

Both design projects identify the various stakeholders in a common platform and provide a detailed depiction of the obstacles and contexts in which any design solution must fit. The Rich Picture and CATWOE Analysis have the potential to be extraordinarily effective in bringing fullness and clarity to the design problem.

7.8   BLACK AND TRANSPARENT BOX

After taking a step back to construct your Rich Picture and perform the CATWOE analysis, now it is time to get back to the nitty gritty of detailed engineering design. A helpful tool at this

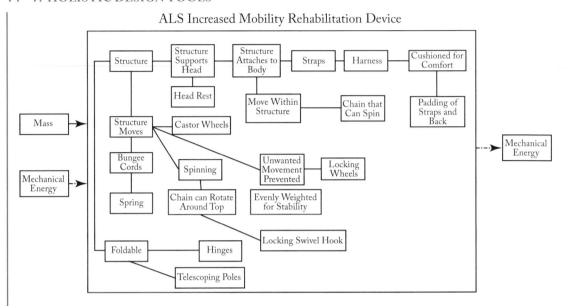

Figure 7.11: ALS enhanced mobility transparent box.

stage of the process involves considering input and output using the black box and transparent box (Dym, 2008 [31]).

In several of your previous engineering science courses, you studied input and output. In this model we model the design project as an open system with flow in and out across the boundary of the design project. We use a **Black Box** to consider only the inputs and outputs and a **Transparent Box** that allows us to look inside and see what is going on.

The combination of black and transparent boxes adds clarity to the beginning and end states of the transformative processes that occur in the design process. While both identify the inputs and outputs for the design, the transparent box allows the design engineer to follow the transformation from raw material to the final product. Examples from the **Waste-to-Wealth** project and the **ALS Enhanced Mobility** project are shown below.

CHAPTER 8

Specific Design Tools: Design for X $\Rightarrow$ Excellence

"Excellence is going the extra mile."

Joyce Meyer

8.1 DESIGN FOR X $\Rightarrow$ EXCELLENCE

In the prior chapter we described a set of holistic design tools that were applicable to all design projects. In this chapter we are offering you another set called **Design for Excellence** tools.

Design for Excellence (DfX) is used extensively in the existing design literature, where the X in "Design for X" represents a targeted aspect of the design. The DfX algorithms address a wide range of issues that the design engineer will need to confront in the successful development of a solution including the following.

- Safety: Identifying the hierarchy of safety hazards and reducing or eliminating as many as possible.

- Reliability: Estimating the reliability of the design elements individually and collectively.

- Cost: Approximating the cost(s) as soon as possible with an intent to limit those cost(s).

- Manufacturability: Ensuring the design can be easily produced.

- Usability: Insuring end users can use the device easily and safely.

- Environment: Analyzing the life cycle of the mechanism to determine the impact of the project on the environment.

- Cultural Sensitivity: Developing an awareness of and respect for cultures that differ from those of the design team members.

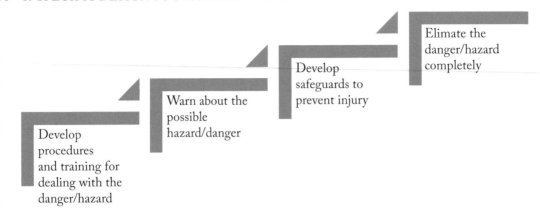

Figure 8.1: Safety hierarchy from least desirable to most desirable.

8.2   DESIGN FOR X ⇒ SAFETY

Safety is the most important consideration in all design projects. It involves identifying the hierarchy of safety hazards and reducing or eliminating as many as possible. Teams must anticipate the potential failures and hazards in a system and propose remediation actions to either prevent or lessen the possible impact of such occurrences.

A general approach to increasing the safety of a design is to:

• minimize the number of hazards and the amount of hazardous materials present at any one time;

• replace any hazardous material with a less hazardous one;

• reduce the strength/occurrence rate of any hazardous material or feature; and

• design out problems rather than adding in additional complexity.

Frequently, when dealing with safety issues, reference is made to a Safety Hierarchy in Fig. 8.1.

One of the most important parameters used to determine safety is referred to as *the factor of safety*. In engineering, a factor of safety is a calculation that determines how much stronger a system is than it needs to be for an intended load. Many systems are purposefully built much stronger than needed for normal usage to allow for emergency situations, unexpected loads, misuse, or degradation. Factors of safety are often estimates based on experience because frequently it is not practical to determine with certainty under what conditions a system will fail.

Factors of safety vary with the engineering application. Required values are specified by various professional societies' publications and by government agencies such as the Occupational Safety and Health Administration (OSHA). OSHA is a federal agency responsible for enforcing the provisions of the *US Occupational Safety and Health Act*. This act includes creation of specific regulations concerning workplace safety and enforcement of those regulations. Another source of information on factors of safety is the government publication, *2010 Americans with Disabilities Act (ADA) Standards for Accessible Design*.

8.3   DESIGN FOR X ⇒ RELIABILITY

Although there are many ways to address reliability, one of the most common is the **Fault Tree Analysis (FTA)**. Using a fault tree diagram, the FTA provides a systematic approach of identifying ways a system might fail and developing the best solutions for minimizing risks.

FTA is regularly used in reliability engineering and safety engineering in industries to mitigate and reduce the hazard. These industries include the aerospace industry, chemical and process industries, the nuclear power industry, the petrochemical industry, and the pharmaceutical industry. It is used as well in software engineering.

A fault tree has three main components:

- a diagram of the entire process;

- a listing of the separate events which make up the process; and

- the way the events are connected in space and time (referred to as "gates"). The various kinds of gates are shown in Fig. 8.2.

This process can help the team not only understand the series of events that can lead to a flaw in the system or device, but it can also help to demonstrate compliance with safety rules and regulations like the *Americans with Disabilities Act*. It can also be a useful tool in the minimization and optimization of the resources.

The Fault Tree constructed for the **Body-on-a-Chip** project is shown in Fig. 8.3.

8.4   DESIGN FOR X ⇒ COST

When you are Designing for Cost (DfC) it is important to:

- create a list as soon as possible in the design process of all raw materials and components that you anticipate needing for your design;

- estimate the cost for each item; and

- estimate the time required to fabricate each sub-assembly in your design. This is particularly important if the work must be done by an outside source.

To reduce the costs, there are several product design techniques available such as Quality Function Deployment, Design for Manufacturing and Assembly, Value Analysis/Function Analysis, Early Supplier Involvement, Activity Based Costing, and Target Costing. These different techniques are described at length in current design literature.

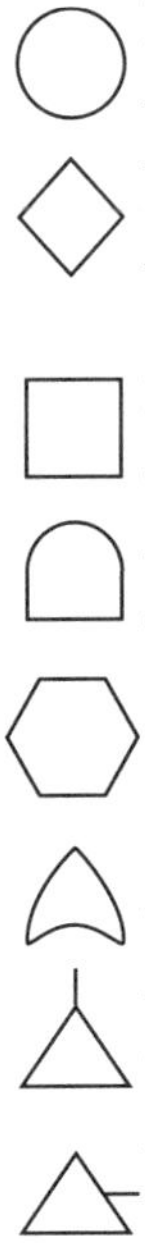

Basic Event
An initiating fault requiring no further development

Undeveloped Event
An event that is not develped further, either because it is considered unnecessary,
or because insufficient information is available

Intermeciate Event
An event arising from the combination of other events

AND Gate
All input events must occur for the output to occur

INHIBIT Gate
Output fault occurs if the single input fault occurs in the presence of an enabling condition

OR Gate
The occurence of one or more input events will cause the output to occur

Transfer In
The tree is developed further elsewhere

Transfer Out
Indicates the place where the development takes place

Figure 8.2: Various gate symbols used in fault tree analysis.

8.5   DESIGN FOR X ⇒ MANUFACTURABILITY

Design for Manufacturability (DfM) is the general engineering practice of designing products in such a way that they are easy to manufacture. While the concept exists in almost all engineering disciplines, the implementation differs widely depending on the manufacturing technology.

DfM will allow potential problems to be fixed in the design phase (which is the least expensive place to address them) rather than the construction phase. In addition to cost, other factors can affect the manufacturability including the type of raw material, the form of the raw material, dimensional tolerances, and secondary processing such as finishing.

8.6   DESIGN FOR X ⇒ USABILITY

Design for Usability (DfU) means designing a system or device that is easy to use, easy to learn, easy to remember (the instructions), and helpful to the user. Gould and Lewis (Gould, 1985 [38]) recommend that designers striving for usability follow these four design principles.

- Early focus on end-users and the tasks they need the system/device to do.

- Empirical measurement using quantitative or qualitative measures.

- An iterative design process with the goal to improve the design after each stage.

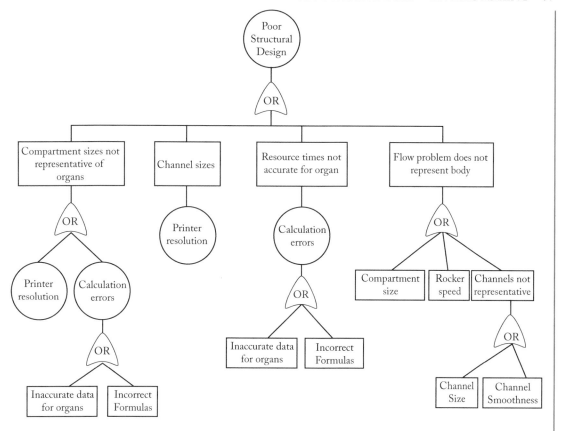

Figure 8.3: Fault tree analysis for body-on-a-chip.

- A user-driven design team in direct contact with potential users.

Evaluation methods like role playing (i.e., team members using the device or process as if they are the user) and surveying (i.e., giving the client a questionnaire based on the Likert scale) can contribute to the team's understanding of potential users and their perceptions of how well the product or process works.

8.7   DESIGN FOR X ⇒ ENVIRONMENT

Every design team must address the question, "What is the proposed design's environmental impact?" **Life Cycle Assessment (LCA)** (Ilgin, 2010 [46]) is a tool that can provide a framework for measuring the impact of a design solution and answering that question.

A Life Cycle Assessment consists of four steps (Fig. 8.4).

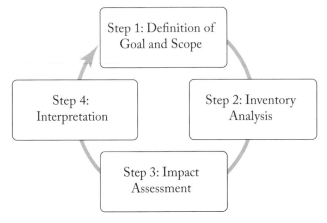

Figure 8.4: Life cycle assessment flow chart.

Step 1: **Definition of Goal and Scope**

Here the extent/limits of the analysis are formally defined. The questions to be addressed are:

> What is to be assessed?
>
> How much of the design will be assessed (functional unit)?
>
> What regulations, or laws, are to be considered?
>
> What will not be included in the assessment?

Step 2: **Inventory Analysis**

This step analyzes the environmental inputs and outputs of a product or service. It is the data collection phase of LCA with the goal being to quantify those environmental inputs and outputs.

> What are the inputs in air, water, energy, and natural resources?
>
> What are the outputs in exhaust gases, wastewater, solid wastes, heat, and other forms of energy?

- ### Step 3. **Impact Assessment**

  This step consists of three key tasks.

  - Task 1: Selection of **indicators** and any relevant models. Some of the more important indicators are
    * global warming potential,
    * acidification of the land potential,

- * eutrophication potential,
- * human toxicity potential,
- * freshwater aquatic eco-toxicity potential, and
- * marine aquatic eco-toxicity potential.

– Task 2: Classification.

Indicators come in different shapes and formats. For example, emissions from harvesting raw materials are very different from the emissions from producing electricity. In this task, we try to group the selected indicators into impact categories. This procedure converts a set of indicators into one category referred to as an **impact category**.

– Task 3: Impact Measurement.

Once we have organized the indicators according to categories, we need some mechanism whereby we can sum up all this different information into one meaningful result.

To make an analogy, suppose we had collected data for a speeding car. While some measure the speed in ft/sec, others measure it in km/hr, and still others measure the speed in furlongs/fortnight. This points to the need for a common unit of measure. In LCA, this is accomplished by creating what is referred to as an **equivalent**. Finding equivalents refers to putting all the results found in the inventory into a common unit … typically kilograms of pollution—which then can be summed and compared to results of other designs.

- Step 4. **Interpretation**

  This final step includes:

  - Identifying significant issues based on inventory phase.
  - Evaluating the study and its completeness.
  - Making conclusions, citing limitations, and proposing recommendations.

There are now several software packages that enable an LCA to be performed. Initially most are free; however, after a few trial runs, there tends to be a nominal charge.

8.8   DESIGN FOR X ⇒ CULTURAL SENSITIVITY

The challenge of becoming culturally sensitive designers requires a positive attitude toward cultural diversity—seeing it as a rich source of inspiration for new concepts and for intercultural connection. Cultural sensitivity in designing is not merely about understanding cultural differences, but more importantly about recognizing cultural dilemmas and finding the common grounds to build upon. Ultimately it is developed through exposure to cultural differences, by working in multi-cultural contexts, and by facing the dilemmas that are raised within those contexts.

To better understand what this means in the context of capstone design, consider this project undertaken a few years ago by students working on a design project for the Onondaga Nation, a Native American nation in upstate New York.

A student design team was asked by one of the Onondaga clan mothers (in this case, the Wolf Clan) to develop a device that would help in the creation of corn soup. In the initial client meeting, the team learned that the Nation is organized in different clans with each clan having a clan mother as well as a chief serving in leadership roles. The clan mothers gather together regularly to supervise the procedures of Nation's ceremonies and oversee the preparation of the foods to be served—the creation of the corn soup being one of their most important ceremonies. (Although the Onondagas were originally primarily hunter gatherers, they had developed their own food system to survive which included corn, beans, and squash—referred to as the "3 sisters.") The corn is first dried, then ground into flour to make a "mush," and finally boiled to make bread. The problem presented to the student design team was that the clan mothers who prepared the "mush" were becoming quite old, and the arduous task of grinding the corn was becoming more and more difficult.

Assuming that time efficiency of the corn grinding process was paramount, the team's immediate response was to suggest a high-speed electric-powered mixer that would turn the kernels into "mush" within seconds. However, this idea was summarily rejected by the clan mothers. They explained that they did not want to remove their connection to the soup making process. They wanted to continue to separate the kernels and stir the soup; they simply needed some help in doing the arduous task of grinding the corn. They also told the team that they wanted a quiet process so that they could work in silence or softly share stories with other members of the community.

The team's final design was a hand-powered crank which used gearing to reduce the force needed to stir the "mush" (Fig. 8.5). It was as quiet as the original process but required much less strength to turn the blades. Although the grinding of the corn took a little less time, the sacredness of the ritual was only slight disturbed.

For this team to be successful, it was necessary to spend time in the Onondaga Nation, visiting with the clan mothers and being introduced to other important elders. It was also was necessary for the students to learn about the history of the Native Peoples in North America, especially the Onondaga since this was the nation of their client. The team members read from various historical texts which offered the traditional recounting of the interactions between the U.S. and Native Peoples as well as views written from the Native people's perspective.

The student team realized that their notion of time and its importance was radically different from that of the Native people. For the clan elders, the time it took to accomplish a task was not considered something to reduce; in their view, a task "simply takes the time it takes." Additionally, the team was surprised to discover that the Onondaga Nation is very much a cash-

Figure 8.5: Mixing blades for preparation of Onondaga corn "Mush."

based society with little use for checks, credit cards, or trust for banks. Again, this was in contrast with the students' perspective.

Another element in this project that often overshadowed the communication between the team and the clan mothers involved the issue of developing trust. It came as a total surprise to the student team that there was, in fact, a deeply buried sense of distrust that needed to be overcome. How could there not be considering the genocide toward Native peoples that has occurred in the United States throughout the 1800s and the discrimination that continues to this day?

CHAPTER 9

Ethics in Design

I can do no other than be reverent before everything that is called life. I can do no other than to have compassion for all that is called life. That is the beginning and the foundation of all ethics.

Albert Schweitzer (Schweitzer [60])

9.1 INTRODUCTION

To meet ABET accreditation requirements, capstone design students are often asked to review the codes of ethics of their appropriate professional societies and/or possibly respond to a case with an ethical dilemma. While both activities hold some merit, it is our belief that you as students need a much more in-depth examination of ethical paradigms in order to acquire sufficient background to carefully consider and respond to the ethical dilemmas that are occurring in our increasingly complex world. For example, one of the most challenging issues for our profession as we move forward in the 21st century and witness the rapid advances in synthetic biology is the question: what is the nature of life?

Ethical theories provide a framework that help us make the difficult choices engineering presents. This chapter includes an overview of some of the traditional ethical theories that our profession has used over the years and offers a new paradigm for the ever-more complex future.

9.2 WHAT IS ETHICS?

Ethics refers to a system of moral principles and perceptions that affect how people make decisions and lead their lives. Ethical theories provide a method or process for deciding how to act in each situation. They serve as guides to help you not only analyze complex problems and issues but also explore what is good or bad, right or wrong—not only for individuals and society, but also for the environment and other species in the natural world.

Because of the complexity of the problems that professional engineers are called upon to resolve, engineering requires extensive ethical training. According to the National Society of Professional Engineers (NSPE Code of Ethics Preamble, 2020 [53]),

> "As members of this profession, engineers are expected to exhibit the highest standards of honesty and integrity.... Accordingly, the services provided by engineers

require honesty, impartiality, fairness, and equity, and must be dedicated to the protection of the public health, safety, and welfare. Engineers must perform under a standard of professional behavior that requires adherence to the highest principles of ethical conduct."

9.3   ETHICAL DILEMMAS

Engineers regularly face what is known as an ethical dilemma. This is a situation in which a difficult choice must be made between two courses of action, both of which entail transgressing a moral principle. In other words, two courses of action are available, but each violates a principle which we would normally follow were it not for the situation in which we find ourselves.

To better understand the nature of an ethical dilemma, consider the following pair of "trolley car cases" (D'Olimpio, 2016 [28]):

Case 1: There is a runaway trolley barreling down the railway tracks. Ahead on the tracks are five people. The trolley is headed straight for them. You are standing some distance off in the train yard, next to a lever. If you pull this lever, the trolley will switch to a different set of tracks. Unfortunately, you notice that there is one person on the side-track. You have two options: (1) Do nothing, and the trolley kills the five people on the main track. (2) Pull the lever, diverting the trolley onto the side-track where it will kill one person.

Pause a moment and consider what you would do.

Case 2: There is a runaway trolley headed toward five people again. Only, this time, you are not in the train yard next to a lever. You are on a bridge, watching the events from above the tracks. There is a very large man next to you. You realize that, if you push him off the bridge and down onto the tracks below, the trolley will hit and kill him, but his body is so large that it will stop the trolley before it reaches the five endangered people. You have two options: (1) Do nothing, and the trolley kills the five people. (2) Push the large man off the bridge, so that he dies, but the five others are saved.

What would you do?

Both cases present us with an ethical dilemma. Strangely, 90% of people say that it is acceptable to kill the one person in Case 1 by pulling the lever, while 95% of people say that it is not acceptable to kill the one person in Case 2—even though the results are the same. In each case, the option is to kill 1 person, or let 5 die.

While you may not work on designing trolley cars as part of your future career in the 21st century, you will likely be confronted with situations which force you to make a choice between two undesirable alternatives.

9.4   A FRAMEWORK FOR ETHICAL DECISION MAKING

This is a classic framework frequently used by engineers confronted with an ethical dilemma. Adapted from the *Universal Traveler* (Koberg, 1981 [43]), it offers a step-by-step approach to making a difficult decision.

1. *Recognize the existence of an ethical dilemma.* Does this decision involve a choice between a good and bad alternative, or perhaps between two "goods" or between two "bads?" Moreover, who decides what is "good" and what is "bad?"

2. *Research and reflect upon what is known and what remains unknown.* What are the relevant facts of the case? Can I learn more about the situation? Do I know enough to decide? (The information will never be complete. Eventually we are forced to fill in those gaps with our own thoughtful reflection.)

3. *Identify the stakeholders.* What individuals and groups should be included? Who decides who is to be included in the considerations?

4. *Brainstorm options for action.* (This is a good place to use another TRIZ diagram!) Have the considerations been limited in any way by the inertia associated with the *status quo*? Have both critical and creative thinking skills been utilized fully?

5. *Evaluate Alternative Actions.* One approach to evaluating the various options - which we will discuss in detail in this chapter—is to consider those options in terms of various ethical paradigms. These include traditional models such as: Utilitarian; Duty-based; Rights-based; Fairness/justice-based; and Virtue-based. An emerging ethical paradigm, the morally deep world view, will also be discussed.

6. *Select and test.* Select from the various options after consideration of the different ethical paradigms and test several options out. Would the decision be the same if roles were reversed? Or would the decision be the same if perhaps the result was published in the ***New York Times***? One that we have found to be very useful is to ask oneself what would my grandmother think of my decision (the *nonna* test)?

7. *Act and reflect on the outcome.* How did my decision turn out and what have I learned from this specific situation? How can I go forward now to the next ethical dilemma? What have I learned about decision-making? How would I do things differently the next time? What do I regret?

If you compare this seven-step method used to address ethical dilemmas with the seven-step method used in solving design problems (Fig. 9.1), you discover that they are in fact the same!

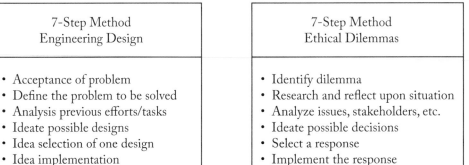

Figure 9.1: Seven-step methods.

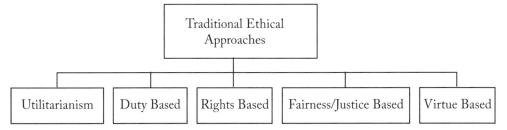

Figure 9.2: Traditional approaches to engineering ethics.

9.5   TRADITIONAL APPROACHES TO ETHICAL DECISIONS

As you can see in Step 5 of the ethical decision-making framework in Fig. 9.2, an ethical theory or paradigm is essential as you make the difficult choices that the engineering process often demands. Here is an overview of five of the more commonly utilized ethical approaches.

The Utilitarian Approach

The "**utilitarian approach**" focuses on outcomes/consequences of actions and asks the question, "Do the majority of people benefit from this outcome?" In this approach the ethical action is viewed as the one that does the greatest good and the least harm for all who are affected; this includes customers, employees, shareholders, the community, as well as the environment.

It involves three steps:

1. identifying the various possible outcomes;

2. determining who will be affected by each action as well as the benefits and harms involved in each outcome; and

3. choosing the action with the greatest benefits and least harms.

In the trolley problem, if one utilizes a "utilitarian perspective," saving the five lives would be the obvious ethical choice.

The Duty-Based Approach

A "**duty-based approach**" is concerned with doing the "right thing" regardless of the consequences of the actions. In this approach, a universal set of moral rules exists that includes: it is wrong to kill innocent people, to steal, and to tell lies while it is right to keep promises. The theory proposes that people have a duty to do the right thing—even it if produces more harm or less good than doing the wrong thing. Duty-based ethics are usually what people are referring to when they say, "It is the principle of the matter."

Applying this approach to the trolley car problem, using the lever to cause the trolley to roll down the track and kill an unsuspecting person is definitely the wrong thing to do—as would be pushing someone off a bridge.

The Rights Approach

The "**rights approach**" is based on the belief that humans have a dignity based on being human and that they possess certain rights, the most important of which is to be treated as "ends" and not merely as "means to an end(s)." In this approach, a human's moral rights include the right to the truth; the right to have autonomy over life's choices; the right to be free from harm, and the right of privacy. Actions that violate the rights of the individual are consider wrong. The question that this approach asks is "Does the action respect the moral rights of everyone involved?" Unfortunately, what quickly becomes apparent is that the rights of one person or group are in conflict with the rights of another.

From a rights perspective, it is not so easy to flip that switch because doing so would make the person a "means to an end" and would knowingly subject the person to harm. Additionally, it would be considered unethical to push the unsuspecting person off the bridge.

The Fairness/Justice Approach

The "**fairness/justice approach**" asks the question "How fair is this action? Does it treat everyone involved equally or does it show favoritism and/or discrimination?" In this approach, an ethical action treats all human beings equally—or, if unequally, then as fairly as is possible based on some standard that is defensible. Difficulty arises in this approach when one attempts to formulate a working definition for "fairness" or a "just action."

Although this approach may not offer much insight into the trolley problem, it reinforces the challenge in defining what is fair and just.

The Virtue Approach

The "**virtue ethics approach**" is a theory that suggests that people should be judged by their character traits, not by specific actions. It suggests that there are certain ideals called virtues toward which we should all strive such as honesty, courage, compassion, generosity, tolerance, love, fidelity, integrity, fairness, self-control, and prudence. An individual who has developed good character traits (virtues) is judged as a morally good person. An individual who has developed bad character traits (vices) is judged as a morally bad person.

This approach, while it may not help with the trolley dilemma, can be very useful with other ethical dilemmas. It encourages us to ask the questions, "What kind of person should I be? Will this action contribute to the development of good character in me? In my community?"

A decision-maker is not limited to using one ethical approach. In fact, it can be quite helpful to utilize more than one ethical paradigm to increase the level of confidence in the rightness of the decision when faced with a complex situation, when the decision will make a significant difference to a person or organization, or when there are contrary points of view.

9.6   ANOTHER ETHICS DILEMMA: THE CASE OF THE OVERCROWDED LIFEBOAT (WEI, 2010 [69])

In the trolley car case we considered two straightforward situations involving personal choice—flipping a switch or pushing someone off a bridge. Let us now consider a more complex case.

> After a shipwreck, you find yourself in a lifeboat that has enough room and provisions for no more than 50 people. However, there are currently 75 people in the boat including yourself—men, women, children, old and young, rich, and poor, passengers and crew members. In addition, there are another 100 people treading water around the boat for a total of 175 people. The people in the water will be overcome by hypothermia in less than 3 hours because the water temperature is less than 50 degrees. Sharks are beginning to circle the area. Since the ship sank in a remote area, the chances of a rescue anytime soon are poor, especially with no other lifeboats available. At this point, you do not even know if the lifeboat's locator beacon system is working or what supplies are stowed on the lifeboat.

> For whatever reasons, the people in the boat, as well as those in the water, have agreed that you should be the one to make any decisions necessary to maximize the chances of survival for the 50 people who can fit in the lifeboat. Everyone will indeed drown if you do not choose only 50 people including yourself. And you must do it in a hurry with no real information about the rest of the passengers except the obvious facts of sex and likely age. Remember, you cannot get out of making the decision; you must choose a total of only 50 people, and you have to do it in a hurry and with the information you have or can get very quickly.

> Pause for a moment and consider what you would do.

Utilitarianism only suggests that you choose a path that does the greatest good and results in the least harm. How does that help you here? You still must eliminate 25 survivors who are in the lifeboat. How would you make that decision? What criterion can you imagine that would be considered ethical? A lottery selection? Gender preference? Children first? Old people last? The survival of the most fit? The survival of those with specific skills? Do you select based on some sense of duty? What is the "right" thing to do? What about the moral rights of everyone? What is fair and/or just? What action on my part would be seen as virtuous?

There are no easy answers. Using a variety of the ethical approaches helps you identify the strengths and weakness of each approach and hopefully gain more insight into what the "best" ethical decision might be.

We believe that there is something missing in each of these approaches. In each of them, it is assumed that there is one person or one individual design team making an ethical decision. As the ethical challenges engineers face in the future become more complex, engineering needs to both consider and involve more people from the "community" in the design process. This is needed in order to gain an expanded view of both the problem and its proposed solution and impact on others far beyond the client. Such an approach is described in the next section of this chapter.

Our Back Page: A Capstone Design Project that Challenged Traditional Approaches to Ethics

In the 1990s while we were on the faculty at another institution, a student design team was tasked with designing and delivering a movie-ticket tearing device for a young man with severe physical limitations. The young man's job entailed welcoming patrons into the movie theater, taking their tickets, separating the halves, and keeping half the ticket for the theater and returning the other half to the patrons. Problems arose because of both the time it took for him to tear the tickets and the strength that was required for accomplishing that task many times in succession.

The student team followed the design process flawlessly and delivered the ticket-tearing device to their client who immediately put the device to use. Soon the project began to garner local and even national attention. The then-mayor of Philadelphia declared a Day of Celebration to be observed on the day the device was first used by the client. The student team was interviewed first by local media and subsequently included in a segment on the CBS Radio Network's *Osgood Files*. Everyone—the client, his family and support staff, the university administration, student design team, sponsoring agency, local community, and faculty advisor—all were delighted with the project and the outcome.

After the completion of the project and its fanfare, the student team went on with their lives, and the young man returned to his prior life. Although he now had the ticket-tearing device that helped him at work, he was greatly affected by the sudden decrease of activity and fanfare. As a result of finding himself no longer the focus of the great attention that he had received throughout the project, within a few weeks he had an emotional breakdown and was subsequently hospitalized.

At this point we found ourselves faced with a situation that did not fit neatly into one of the traditional ethical approaches. The design team had met all the conditions placed upon them by professional codes, their client(s), and others. While the project seemed a great success initially, there is no doubt that the fanfare played a significant role in the heartbreaking setback the client experienced after the project's completion.

The entire episode forced us to deeply reflect and search for an ethical code which would reduce the possibility of another tragic occurrence. Some wide-ranging questions arose as we reflected on our experience. Would the outcome have been different if we had consulted with other professionals such as a counselor or social worker? Why did we assume that the responsibility of the design team ended with delivery of the device?

We also began to question what are our responsibilities toward a client not only in the present but also in the future? What are our responsibilities to the local community? To other individuals with disabilities? Should we even have taken the project on at all?

Such questions did not arise while following the traditional approaches. After much soul searching, it was clear to us that a new approach to engineering ethics was needed to address complex design problems like this one. As a result, "a morally deep world view" came into being.

9.7   AN EMERGING NEW ETHICAL PARADIGM: THE MORALLY DEEP WORLD VIEW

A **morally deep world approach** to engineering ethics (Catalano, G. D., 2014 [19]) is derived from Lawrence Johnson's work, *A Morally Deep World* (Johnson, 1989 [42]). Johnson contends that we should reject purely human-centered approaches to the environment and recognize that all living things (and some ecological wholes, such as species, ecosystems, and the biosphere as a whole) have intrinsic value and deserve moral consideration and respect. His work emphasizes the importance of recognizing and considering not only an individual being but also the community in which that being exists.

For example, in writing about predator-prey studies involving lions and deer, he stated that a morally deep world view would consider the impact of an ethical decision on not only both the lions and deer (i.e., individually and as a species) but also the ecosystem in which they exist.

A morally deep world approach to engineering ethics then is an ethical approach that is not only rooted in both respect for the individual (i.e., each of us as a person) as well as the community (i.e., our family, our country, our planet), but also one that seeks the wisdom of the many different voices and cultures across the globe. Included in those voices are the voices of the other species with whom we share this planet. The morally deep approach to engineering ethics is one that calls for an expansion of our responsibilities as a profession in the same way the ripples move out from the impact of a stone on a still pond.

It asks the following questions.

- Who individually is impacted and how, now and in the future?

- What communities are impacted and how, now and in the future?

- What cultures are impacted and how, now and in the future?

- What other species are impacted and how, now and in the future?

- What ecosystems are impacted and how, now and in the future?

It includes the following two principles.

1. When confronted with an ethical dilemma, both the individual and the communities in which that individual is a part must be considered. Both matter ethically. It is not an "either/or" choice; rather it is more inclusive—considering all who might be impacted by the decision.

Consider the following real-world design project in which I was a participant. The project, undertaken by the U.S. Air Force, involved modeling the diffusion of high-powered laser weapons in the lower regions of the Earth's atmosphere. It was discovered that turbulence in the atmosphere was diminishing the effectiveness of the beam as a weapon. Although most of the testing was done over the desert regions of New Mexico and Arizona, away from populated areas, considerable damage was done to the local animal populations and ecosystems. When Native tribes and environmental groups brought media attention to this issue, the Air Force initially was adamant about the need for the tests. Eventually after much debate and the refusal of several engineers to continue working on the design project because of the damage to the local ecosystem and its inhabitants, the work was halted.

2. It encourages the participation of not only the client but also a wider group of people members to contribute to the discussion—to share their expertise, their different perspectives, etc. In this approach these members are essential to the whole design experience and are referred to as the "integral community."

An issue that arose for the design team's **Waste-to-Wealth** project can provide for further insight into what this second principle means. The project's goal was to produce usable plastic sheeting from plastic trash bags and bottle that were spread across the countryside during the flooding that occurred in the rainy season. The hope was that the plastic sheeting could be sold for a profit which would encourage more recycling and thereby reduce damage to the environment. It seemed a "win-win" result. However, during the project, the team discovered that a

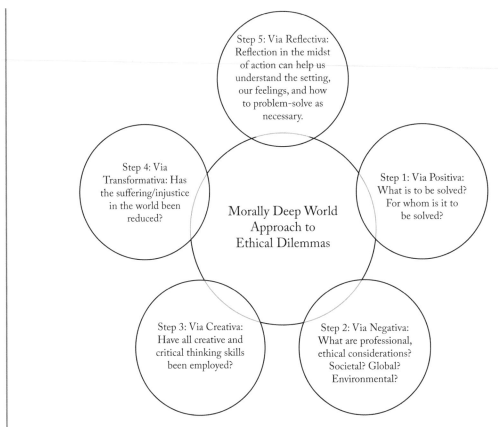

Figure 9.3: A new ethical approach using morally deep world view.

different and significantly more impoverished segment of a nearby community already had a recycling effort in place. They realized that their project would ultimately adversely affect that nearby effort. After seeking the wisdom of many different local officials, the team decided to go ahead with the project—though not without considerable soul searching and reflection. In this situation, the "integral community" included not only the design team and their clients but also a wide variety of officials and citizens in the community.

The basic framework to this approach has five steps (see Fig. 9.3).

> *Step 1: Via Positiva.* The problem is identified, fully accepted, and broken down into its various components using the vast array of creative and critical thinking techniques which engineers possess. What is to be solved? For whom is it to be solved?

> "If we had utilized this approach in the case of the ticket-tearing device, we would have considered the young man's family, his support staff, and others

with whom he interacts. We would have taken the time to investigate more about his condition and the effects his disabilities have on his state-of-mind."

Step 2: Via Negativa. Reflection on the possible implications and consequences for any proposed solution are explored. What are the ethical considerations involved? The societal implications? The global consequences? The effects on the natural environment?

> "In the case of the ticket-tearing device, we would have considered what the future might hold for this young man as he returned to his life apart from the media attention."

Step 3: Via Creativa. The third step refers to the act of creation. The solution is chosen from a host of possibilities, created, and then evaluated as to its effectiveness in meeting the desired goals and fulfilling the specified criteria.

Step 4: Via Transfomativa. The fourth step asks the following questions of the engineer: Has the suffering in the world been reduced? Have the social injustices that pervade our global village been even slightly ameliorated? Has the notion of a community of interests been expanded? Borrowing from the Greek poet Aeschylus, is the world a kinder, gentler place?

> In the case of the ticket-tearing device, we would have been forced to consider whether the design project should be built at all. We might have asked, "Will he be better off because of the project? Will his family be better off?"

Step 5: Via Reflectiva. The fifth and final step may be the most important step of all as it forces a reflection upon the action(s). Here we ask the question what have I learned from this entire experience?

Many design students tend to feel unsettled by this case, saying "We cannot possibly know all of the possible consequences to a design." That is true; there are no "approved solutions" to complex ethical situations. However, we believe that it is essential to take a wider more expanded view of each situation and consider ALL of the possible issues and consequences—realizing that while there are no easy answers to these questions, the questions are important in and of themselves.

9.8 COMPLEX ETHICAL CHALLENGES CONFRONTING ENGINEERING IN THE 21ST CENTURY

A recurring theme throughout this text is the fact that problems that engineers are facing now and will face in the future are becoming more and more complex. This is especially true for ethical dilemmas. Now that you have been introduced to a new ethical approach, the morally deep world view, consider two of the complex issues that confront us today—the health of the

Earth and the global pandemic of 2020. While on the surface they appear to be quite different phenomena, both raise similar issues and suggest the relevance of the morally deep world view.

Health of the Earth

Environmental issues are related to our species' impact on the global ecosystem which includes the living world of plants and animals, habitats, land use, and natural resources. As we have seen earlier in this text, while green engineering asks us to account for the environmental impact of our design solutions, it does not offer guidance when we are confronted with ethical issues that arise.

Let us look at some of the "green" ethical issues facing us today.

- **Air Quality**: air pollution, ozone pollution, ties to human health with asthma, diesel emissions, etc.

  The Earth has one atmosphere and we all are affected by the quality of that atmosphere. For example, the pollution that occurs due to coal-fired electrical plants in Asia directly affects the quality of the air we breathe—wherever we live. This poisoning of the air poses health hazards particularly for those whose breathing is already compromised. How do we address air quality issues? Locally? Globally? Whose perspectives are included?

- **Biodiversity**: conservation of biological diversity, and the Sixth Great Extinction event.

  Our planet is in the midst of the Sixth Great Extinction. Species are disappearing more rapidly than ever before, and their loss greatly impacts us as humans. For example, bees are threatened with extinction due to the overuse of pesticides, and yet they are responsible for the pollination of plants and ultimately the production of much of our food. With fewer bees, food shortages will inevitably result. What expertise do we need to address these issues? How long into the future might we look?

- **Climate Change**: "global warming," greenhouse effect, loss of glaciers, climate refugees, climate justice, equity, etc.

  Global warming is greatly affecting the already shrinking glaciers in the Arctic and the Antarctic. The increasingly rapid melting of these glaciers is having catastrophic impacts on the planet including shortage of freshwater, excessive flooding, extinction of animals, disappearance of coral reefs, and a return of lethal diseases that are trapped in the melting ice. How can we address such global issues? Who will speak for the animals and the coral reefs?

- **Pollution**: air, freshwater and ocean pollution, "Great Pacific Garbage Patch," river, and lake pollution, land, toxins, light, point source and non-point source, use of coal/gas/etc., reclaimed land issues.

The devastating impact of ocean pollution includes effects of toxic wastes on marine animals, disruption of coral reef generation and health, depletion of oxygen content in water, failure in the reproductive system of sea animals, effects on food chain, and effects on human health. How will we safeguard marine life? Whose perspectives need to be included?

- **Waste**: landfills, recycling, incineration, various types of waste produced from human activities, etc.

  Landfills dot the U.S. countryside as well as other parts of the world. Organic matter in landfills decomposes and produces greenhouse gases like methane and carbon dioxide. Incinerating waste in landfills and using landfill waste for energy conversion also produces greenhouse gases and air pollution. Other effects of landfills include impacts on biodiversity, pollution of groundwater, and reduction of soil fertility. Who decides the fate of local landfills? Incinerators? Whose perspective matters?

In the 1960s, the *Whole Earth Catalog* (Brand, 1971 [13]) likened our planet to a spaceship, "Lifeboat Earth," sailing through space with the crew working together for the common good. More recently, however, due to the impact humans are having on the planet, our planet has been described as a "Lifeboat with a Leak" (Feffer, 2019 [34]).

Just as on the overcrowded lifeboat we considered earlier, time and resources are similarly limited on "Lifeboat Earth." According to most climate scientists in 2020, the window of opportunity to prevent irrevocable damage is about a decade. Opinion is divided, however, on how to address this problem with the urgency it requires. It is this shrinking window of opportunity that desperately calls out for new engineering solutions which, if they cannot eliminate the leak, at the very least can buy more time for more creative solutions in the future. This shrinking window of opportunity also demands an ethical approach that is not only rooted in both respect for the individual (i.e., each of us as a person) as well as the community (i.e., our family, our country, our planet), but also one that seeks the wisdom of the many different voices and cultures across the globe. We believe a morally deep world approach allows us as engineers to accept and face the complex challenges the 21st century presents.

The Global Pandemic of 2020

As we write this book, the COVID-19 pandemic is bringing into focus many of the same issues as the "Lifeboat Earth" example while tragically introducing several new issues. During the exponential growth phase of the virus, there has been a severe shortage of masks and protective personal equipment (PPE) plus the possibility of a shortage of hospital beds and ventilators (Jeffrey, 2020 [41]). This lack of resources (actual and anticipated) requires that we address issues such as the following.

- **Resource allocation**: If ventilators, hospital beds, and masks continue to be in short supply, and available beds in ICU are also limited, how do we determine who has access to resources like ventilators, hospital beds, and masks? How are priorities established?

- **Personal Freedoms and Privacy Rights**: While several states have mandated mask wearing and social distancing at various times during the pandemic, there has been resistance by some to these mandates. Are personal freedoms and privacy rights being overridden in a pandemic situation? How should questions about mask and self-quarantine requirements be decided and by whom?

- **Fund allocation**: As the world population continues to increase exponentially, government funding for health programs worldwide has come under greater stress during the pandemic. What should be the priority of government funding for health programs? Because we live in a global community, should funding go primarily to organizations like the World Health Organization (WHO), or should money go to the local production, transportation, and distribution of medical equipment? Who decides where it should be distributed?

- **Neighbors**: Not only have different regions of the U.S. experienced different levels of outbreak of the pandemic, different parts of the globe have also been devastated by widely varying infection rates. Should we donate to individuals in need in our local communities or to organizations that support people globally? Should funding priority be given to long-term questions such as the re-igniting of the pandemic or to short-term immediate concerns?

How can we decide who will get the masks, the protective personal equipment (PPE), the intensive care unit (ICU) beds, and the breathing machines? Who should get the funding? What priorities in funding allocation should take precedent? Who will make those decisions? As we discussed with the lifeboat case, what criteria can you imagine that would be considered ethical? A lottery selection? Gender preference? Children first? Old people first? last? The survival of the most fit? The survival of those with specific skills. Do you select based on some sense of duty? Rights? Fairness? Virtue?

We believe that a morally deep world approach to such complex ethical dilemmas holds the most promise for identifying and developing solutions that will work. It is an approach that requires the active participation of a wider community and welcomes a diversity of views. Most importantly, it forces a consideration of both the short- and long-term consequences.

Our Back Pages: Mitákuye Oyás'i? (Lakota Prayer, Grandmothers Council)

We had the privilege of working a brief time with the Lakota People in Pine Ridge, South Dakota. Though our time there was short, the wisdom that they offer to us is timeless and seems particularly relevant for this chapter on ethics. We offer this gift to all our readers.

Figure 9.4: Lakota flag.

Aho Mitakuye Oyasin… All my relations.

I honor you in this circle of life with me today. I am grateful for this opportunity to acknowledge you in this prayer…

To the Creator, for the ultimate gift of life, I thank you.

To the mineral nation that has built and maintained my bones and all foundations of life experience, I thank you.

To the plant nation that sustains my organs and body and gives me healing herbs for sickness, I thank you.

To the animal nation that feeds me from your own flesh and offers your loyal companionship in this walk of life, I thank you.

To the human nation that shares my path as a soul upon the sacred wheel of Earthly life, I thank you.

To the Spirit nation that guides me invisibly through the ups and downs of life and for carrying the torch of light through the Ages. I thank you.

To the Four Winds of Change and Growth, I thank you.

You are all my relations, my relatives, without whom I would not live. We are in the circle of life together, co-existing, co-dependent, co-creating our destiny. One, not more important than the other. One nation evolving from the other and yet each dependent upon the one above and the one below. All of us a part of the Great Mystery.

Thank you for this Life.

C H A P T E R   10

Testing, Testing, and More Testing

Manufacturing is more than just putting parts together. It's coming up with ideas, testing principles and perfecting the engineering, as well as final assembly.

James Dyson

10.1　INTRODUCTION TO THE DESIGN OF EXPERIMENTS

Testing is an essential part of the engineering design process. Testing allows the design team to determine whether the design objectives have been met.

Often student teams complete the prototype of their design and then realize they have not allowed enough time to adequately test the product or process. Lack of testing diminishes the quality of the project and the credibility of the effort. It is a disappointing outcome for team, advisor and client alike. To avoid falling into the trap of "not enough time to test," student teams need a formal methodology that has proven useful in evaluating and assessing design projects. That methodology is referred to as the **Design of Experiments (DoE)** (Kuehl, 1999 [44]).

10.2　DESIGN OF EXPERIMENTS (DOE)

The good news is that DoE is a technique focused upon generating the required data with the minimum of number of experiments or tests run. It is an eight-step process as depicted in Fig. 10.1.

The key to the DoE methodology is to develop a good test plan before entering the testing phase. This test plan should include the following elements.

- A description of the test that clearly defines its scope, goals, purpose, and all-important parameters.

- A listing of the date and location of all tests and names of all participating members.

- Documentation of the actual testing method and testing equipment.

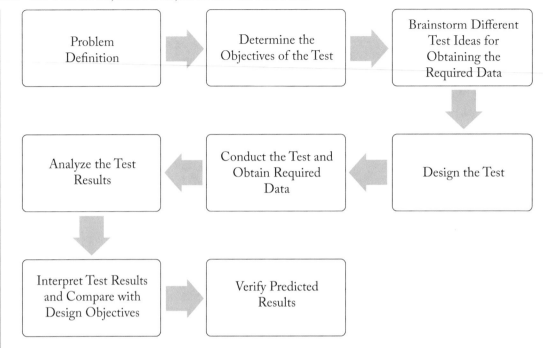

Figure 10.1: Step by step approach to design of experiments.

- Identification of the dependent and independent variables as well as the number of tests and the procedures followed in each.

- Description of the safety protocols.

- Description of the data collection methods used.

10.3   A STEP-BY-STEP APPROACH

While there are a wide variety of suggested approaches that outline steps to complete a DoE, the following generic steps are recommended as a starting point for assessing the final project.

1. Determine the acceptance criteria needed (i.e., confidence level) for determining what will be accepted as passing criteria.

    In statistics, the confidence level indicates the probability with which the estimation of a statistical parameter (e.g., an arithmetic mean) in a sample survey is also true for the entire population. The most used confidence level is 95%.

2. Select two or three variables to be tested.

3. Pick two different test levels for each of the parameters selected (i.e., low/high, on/off, etc.).

4. Determine the number of data points per test. Be sure to answer the question, "What should the sample size (N) be to get a statistically reliable measurement?" Consider a range of different statistical properties from means, medians, and standard deviations to correlations and spectral analysis.

5. Randomize the order to the extent possible.

6. Run the experiment/test and collect data.

7. Keep all other possible control factors as constant as possible as these may affect the validity of the conclusions.

8. Record all factors that may have influenced the test results.

9. Analyze the data. Be prepared to answer the question, "So what?" It is not sufficient to have reams and reams of data, countless graphs, and thousands of lines of code without any effort to interpret the results.

10. Confirm the testing results by either performing the test a second time or designing a new experiment as a test before fully accepting any results.

CHAPTER 11

Obstacles to Success: What Can Go Wrong

If you can find a path with no obstacles, it probably doesn't lead anywhere.

Frank A. Clark (Frank [35])

11.1 EXPECT OBSTACLES

Although there are an infinite number of difficulties that a team might encounter throughout the design process, we have identified four possible obstacles that seem to be among the most common reported by design students. Upon encountering these obstacles, many students discover that they need to develop new skills—including stress management, overcoming procrastination, conflict resolution, and communication skills.

11.2 STRESS

Our Back Page: Why is senior capstone design so stressful? A listing of possible answers includes the following.

- Back in Chapter 1 we referred to Bloom's Taxonomy which illustrates the point that this course requires the highest level of analytical thinking and processing skills.

- Students are asked to solve a problem that has not been solved before. This requires not only analytical skills but creativity. As we saw in the section on the history of engineering, most of the original "engineers" were artists.

- Constant deadlines exist from start to finish.

- Rather than individual coursework which has been the standard for most of the undergraduate experience, students are part of team that must learn to work together seamlessly.

- Students must meet the requirements of not only the professor but also of their client.

- Senior design is often only one of several courses in a given semester.

This is a time of major transition in the life of an engineering student. For many, the capstone design course is the bridge from being a college student to being a professional—signaling a shift from grades to a paycheck. For others it is a stepping-stone into graduate school. As a result, students encounter many sources of stress during this time of change and transformation.

Stress is what we experience when we perceive a demand being made on us—a demand to do something, to be a certain way, to respond to something, or to accept something difficult. This perception of demand, in turn, not only causes our body to respond in physiological ways but is also reflected in our thoughts (inner dialogue), as well as our emotions. When asked, "How does stress make you feel?", many design students report feeling overwhelmed, tense, anxious, nervous, angry or irritable, restless, unmotivated and/or unfocused. Some students report trouble sleeping or sleeping too much, problems with concentration, and/or constant worry.

Good Stress vs. Distress

Because life is filled with demands (real and imagined), stress is a normal part of life; however, not all stress is not detrimental. In certain situations, stress can actually be beneficial. Called "eustress" ("*eu*" means good), it is what we tend to experience when we embark on something new and/or exciting. Remember when you first arrived at the University? You were probably both excited and yet well aware that you were being presented with new challenges in a setting filled with myriad unknowns. This kind of stress can help a person be alert, feel motivated, and prepared to avoid danger; it can also serve to increase and enhance one's performance during a presentation, in a testing situation, or an athletic performance.

Usually when we speak of stress, however, we are referring to negative stress called "distress." This kind of stress results when we feel out of our comfort zone. Feelings of overwhelm and anxiety frequently occur when we believe that we have exceeded our physical, emotional, or psychological ability to cope with a situation.

External vs. Internal Causes

When asked about their stressors, most students can easily name several "demands" that contribute to their sense of distress—including a heavy workload, deadlines, and relationship issues with team, professor/advisor, friends, or family. These are considered *external causes* of stress because they originate from outside the individual.

Many students are not aware, however, that many of the things that we consider stressful arise from within. Stress can result from our thoughts—especially from self-talk that is negative and/or self-critical. Unrealistic expectations, all or nothing thinking, and inability to accept uncertainty can also contribute to stress. In addition, students often experience stress from within due to the lack or deficiency of certain skills which can be easily learned or updated. Time management is a common area where gaining more knowledge and simple tools can relieve one's sense of tension and stress.

Some Tips for Managing Stress

Over the years, capstone design students report that the following stress management tips are useful to them—not only as students but also as individuals.

1. Identify not only the external sources of stress in your current life, but, more importantly, the internal sources.

   The well-known serenity prayer by Reinhold Niebuhr emphasizes that you need a quiet peaceful mind to accept the things you cannot change (i.e., external stressors); courage to change the things you can (internal stressors); and wisdom to know the difference between what you can change and what you cannot change. For example, although you may try, you cannot change another person's character, behavior, or attitudes; you can only change your own—and then, only if you choose to.

2. Develop new life-skills and/or expand your current skill base by learning more about time management strategies for professionals, relaxation skills, communication skills, meditation, and mindfulness.

3. Recognize and address your body's need for adequate sleep, healthy nutrition, and exercise.

4. Become aware of your "inner dialogue" (more about this in the following section) and how it affects not only your stress level but also your behavior.

11.3 PROCRASTINATION

Procrastination is one of the most common obstacles to student success. Procrastination is the habitual and intentional delay of starting or completing a task—despite its negative consequence. It is avoiding or ignoring an unpleasant or difficult task and doing something easier and/or more fun.

This self-inflicted behavior of postponing important tasks and then regretting the decision later when faced with the consequences is a common way that students sabotage their own success because it creates a wide variety of problems in daily life and can ultimately short-circuit one's life goals.

Procrastination is not the same as laziness as laziness suggests passivity, waiting for someone else to step up and handle a situation, apathy, or an unwillingness to act. Procrastination is an active process: one *chooses* to do something other than the task that needs to be done.

Figure 11.1: Three ego states of transactional analysis.

If procrastination is a choice, who is choosing? One model that examines an individual's "inner dialogue" and how it impacts behavior uses Eric Berne's psychological theory called "transactional analysis" as a framework (Berne, 1996 [10]). This model proposes that in all of us there are three aspects of personality that strongly influence how we think, behave, and feel (Fig. 11.1).

The "child" is the part of our personality that is driven by our immediate needs and emotions. In addition to wanting to have things "my way," it just wants to play and have fun with no consideration of possible consequences. An example of the child voice is, "I don't' want to read this text right now. Checking my messages is much more fun."

The "adult" is the part of our personality that appraises reality objectively. Equipped with knowledge, skills, and an ability to make decisions, this inner voice understands the difference between short-term gratification vs. long-term consequences. An example of the adult voice is, "I'd really like to go out with my friends tonight, but I need to complete this task to meet my team's deadline tomorrow morning."

The "inner critic" is the part of our personality that is self-critical and judgmental. While it often results from earlier life experiences with parents or primary caretakers, it can also come from interactions through the years with peers, siblings, and influential adults. Some of the favorite words of this voice are "should," "always," and "never." An example of this self-critic is, "I always screw up …I never do anything right …How could I be so dumb? I can't believe I put that team report off. I should have known better. I am so stupid." Because your thoughts greatly influence how you feel and behave, this destructive voice can erode your self-confidence and self-esteem and cause you to question your self-worth.

The first and most important step in overcoming procrastination as well as other problematic behaviors is to recognize which part of your inner dialogue is driving your behavior at any given moment. If it is not the adult, acknowledge that. Simply being aware of "who is driving the bus" can help you take charge and choose your behavior more wisely.

Your inner dialogue can either help you become successful in all the various aspects of your life, or it can prevent you from reaching your full potential. Because becoming a successful professional requires that you leave behind behaviors that don't support success while at the same time fully developing behaviors that do, making wise decisions from the "adult" perspective makes a great deal of sense.

11.4   CONFLICT

The Nature of Conflict

It is inevitable that sometime during the design effort, conflicts will arise. This happens because of the many differences that exist between people. Team members—as well as clients—have had different life experiences because they come from diverse educational, social, ethnic, and/or cultural backgrounds that result in them seeing the world through different lenses. How team members react to these differences and how conflict is managed will have a large impact of the success or failure of the project.

Not all conflict is created equally. Some conflicts can be constructive and helpful, while others can be destructive and defeating.

- A *constructive conflict* is one in which a solution to a conflict is found that is acceptable to all parties involved. The differing interests and goals of all parties are considered in an effort to maximize opportunities for mutual gains. Collaboration and a willingness to placing "solving the problem" ahead of "getting my way" are essential ingredients of reaching what Covey calls a "win-win" situation: one that is both accepted by all involved and results in the successful end product of the design process (Covey, 2013 [24]). Healthy and constructive conflict is an essential component of high-functioning teams.

- *Destructive conflicts* occur when disagreements gravitate toward antagonism instead of resolution. They have the potential to degenerate to the point that all parties in the conflict lose sight of the overall goal of the project and instead transform their purposes to getting even, retaliating, or demeaning the other person or party. In this situation, which Covey calls "lose-lose," not only is no one is satisfied with the outcome, but possible gains are not realized (Covey, 2013 [24]). The negative atmosphere that lingers at the end of one conflict episode is carried over to the beginning of the next conflict—creating a degenerating or negative spiral. These kinds of conflicts are more likely to occur when team behaviors arise from rigid, competitive perspectives.

A Guide for Constructive Conflict Resolution

This seven-step approach (which is similar to one that we presented for the design process and again for solving ethical dilemmas) can be used to resolve conflict—not only in teams but also in life situations.

1. Acceptance of the Problem

    - Nothing can be resolved until the team acknowledges that a conflict exists.
    - Explore, as a team, the possible impacts of this conflict or disagreement on the team's mission.

- Agree to communicate openly and honestly.

2. Definition

   - Clarify and carefully define conflicting positions of all parties.

3. Analysis

   - Identify and explore the underlying values associated with the different positions.

4. Ideation or Brainstorming

   - After carefully listening to all positions, work together to develop a range of possible resolutions.

5. Idea Selection

   - Using a consensus approach, arrive at a possible resolution.

6. Idea Implementation

   - Implement the chosen resolution.

7. Reflection of resolution

   - As a team, discuss both the resolution and the process used to arrive at consensus in an effort to learn from the experience. This will help to make this a constructive conflict.

Note that the basis of this approach to constructive conflict resolution is gaining understanding of the different perspectives that exist and using that understanding to expand one's own thoughts and beliefs about the issue.

11.5   COMMUNICATION

Effective communication is a crucial key to a design team's success. It allows you as an individual to understand and be understood by others. Communication includes not only your words but also the tone of your voice, the intention behind your words as well as your body language (facial expressions, gestures, and postures).

Four Styles of Communication

While each person has a unique communication style, there are four basic styles of communication that individuals habitually use. These styles are fluid—shifting and changing as one grows and learns from life experiences. Awareness of these different styles can greatly improve your relationships—not only within the team but also in everyday life.

- **Passive**: Individuals who use this style generally fail to express or share their feelings, needs, ideas, or opinions. They often display lack of eye contact, poor body posture, and an inability to say no. Although they are usually easy to get along with, their failure to express their thoughts and emotions can result in miscommunication and built-up anger or resentment.

  > "I don't know …I don't really care …whatever you want to do …"
  > "I'm okay with whatever you want to do. You decide."

- **Aggressive**: The aggressive communicator tends to dominate the conversation, often issuing demands and failing to listen to others. This individual tends to talks over other people, interrupts frequently, is controlling others, points fingers (either physically or verbally by using "you" statements), stares or glares, frowns, and often blames, intimidates, criticizes, threatens, or verbally attacks. The frequent message is "I am right and you are wrong."

  > "You did it wrong! It's your fault. We are way behind schedule."
  > "I can't believe you screwed this up!"
  > "Do you ever do anything right?"
  > "Why? Because I say so …"

- **Passive aggressive**: A person with this style appears passive on the surface but is actually angry or resentful on the inside. They may use frequent sarcasm, make faces, or give someone the "silent treatment." Some mutter under their breath rather than confronting a person or an issue. Spreading rumors or gossip, and sabotaging efforts of others are other ways the anger and resentment may display.

  > "Sure, we can do it your way" and then muttering that your way is "stupid."
  > "Fine… whatever …"
  > After saying something rude, sarcastic, or hurtful, saying "I was only joking."

- **Assertive**: This style is considered the most effective form of communicating; it is open, honest, and direct without being overbearing. An individual with this style expresses needs, wants, desires, ideas, and opinions in a thoughtful and considerate manner and encourages others to speak as well. Maintaining eye contact and having the ability to say no when necessary are other hallmarks of this style. Rather than using aggressive statements beginning with "you" (i.e., You are always late), an individual with this style uses "I" statements like "I feel ____ when ___ occurs."

  > "I feel frustrated when you are late."
  > "This is only my opinion. What are your thoughts?"
  > "I feel hurt when my ideas are attacked so critically. Can we talk about that?"

> "I feel slighted when I am left out of team decisions. I would like talk about that as a group."

Effective communication does not happen overnight. It is a set of skills that can be learned and improved. In addition to understanding the various communication styles, here are some other ways to improve your communication skills and build a stronger team.

1. Listen actively.

   Listen twice as much as you speak; be alert and interested in the speaker; refrain from interrupting.

2. Paraphrase back what you have heard.

   Give feedback to the speaker to let them know that you heard them and that you understand. A verbal example might be "So what I am hearing you say is that you are worried about us not finishing the project on time …Is that right?" In larger contexts like a large class setting, simply nodding your head can indicate that you are engaged with the speaker.

3. Show respect for the speaker.

   Don't' multitask during work sessions; this means staying off of your phone! Make eye contact and use people's names. Most importantly, don't judge or criticize the other person.

4. Be aware of your body language.

   We communicate with not only words but also with our tone of voice, facial expressions, gestures, and body posture like crossed arms and/or legs or tight rounded shoulders. An open friendly body posture sends the message that you are open to hearing what the speaker has to say.

5. Be positive.

   Oprah Winfrey once said, "The greatest discovery of all time is that a person can change his future by changing his attitude."

6. Recognize each team member's strengths and honor it.

   While some team members may be analytical and focus only on facts, others may be skillful at grasping the "big picture" at any moment without overanalyzing. Some, called functional communicators, are great at breaking down situations to consider every aspect, while others have a more personal style allowing them to connect with others and form strong working relationships within the team.

CHAPTER 12

Final Report

"The difference between the almost right word and the right word is really a large matter. 'tis the difference between the lightning bug and the lightning."

Mark Twain (Twain, 2004 [66])

12.1 INTRODUCTION

An **Engineering Project Report (EPR)** is a detailed report explaining the results of an engineering project. Such reports are commonly used by project engineers to keep their companies and their superiors up to date about projects. While memos are short reports that are frequently given as updates throughout the progression of a project, EPRs are usually given at the conclusion of a project and provide a detailed, thorough review of the project including the final details.

Because this kind of report is used extensively in throughout the field of engineering, students typically learn how to write an EPR in introductory engineering courses. The format of an engineering project report can vary somewhat based on who is setting the standards, who is writing it, and who will read it. Because an EPR frequently passes over numerous desks and is read by many people, it must have an easy to understand format and be logically organized so that each reader can quickly find the information that he or she is looking for. This is especially imperative in long projects when reports can be composed of hundreds of pages of compiled data.

12.2 STRUCTURE OF FINAL REPORT

One recommended format for the capstone design report is shown below.

- A **title page** should be used with full identification of names and dates.

- If the report is long, a **table of contents** should follow the title page.

- The **abstract** summarizes the major points in the report in a concise manner, allowing the reader to decide on whether to read the full paper. The first sentence should state what was accomplished. The abstract is not a condensation of the entire paper, but rather a clear statement of the project scope, results achieved, and the conclusions and recommendations drawn from the results.

- An **introduction** indicates the background of the project and the reasons for undertaking it. Some information on previous work is usually included.

- In the **theory and analysis** section, pertinent principles, laws, and equations are stated, and unfamiliar terms defined. Analytical diagrams such as theoretical cycles or flow and field patterns are shown here. Be sure to include all necessary supporting theory without padding with unnecessary filler (i.e., deadwood).

- The **experimental procedures** section describes apparatus and materials. Instrument types, ranges, and identification numbers are indicated. A sketch of the test setup showing relative positions, connections, and flows should also be included.

- The **experimental methods** section describes preliminary results, equalizing periods, duration of runs, and frequency of readings. Special precautions for obtaining accuracy and for controlling conditions should be described. Conformity with or divergence from standard test codes or procedures should be clearly stated.

- The **results** section summarizes the important findings with supporting tables, graphs, and figures. Original data or extensive data tables should be included in appendices. Graphical representation is very important in conveying quantitative results. The use of logarithmic or other special scales should be considered. Deviations from smooth curves should be carefully checked. Apparent discrepancies should be pointed out and explained.

- The **discussion** describes the accuracy and importance of the results. Sources of measurement error should be evaluated. Results should be critically compared with theory, and differences greater than the experimental errors should be explained. Limitations of the theory and tolerances in engineering values should be addressed.

- The **conclusion** section summarizes the conclusions which have been drawn. Conclusions should be supported by specific references to data and results, quoting numerical values, and guiding the reader from facts to conclusions. Conclusions should follow directly from the numerical results quoted without the need for mental arithmetic by the reader.

- **Recommendations** are often more important than conclusions as few experimental projects are an end in themselves. Either the results are to be used for a purpose, or the experimenter sees more work that could be done.

- The **bibliography** lists sources which were directly referenced in the report. Other general references may also be given. Numbered footnotes, or preferably endnotes, are used to list sources in the order of reference.

- Always include an **acknowledgment** section, naming and acknowledging all other contributors to the work as well as people who have contributed ideas or materials. Also acknowledge the project's sources of financial support.

12.3   THE FINAL TEAM PRESENTATION

Presentations generally are either formal or informal. Make sure that you know exactly what is expected of your team in the final presentation.

- Informal presentations generally involve casual clothes, casual posture, and a conversational tone using everyday language. PowerPoint slides are optional.

- Formal presentations require professional clothes and appearance, professional posture, and a less conversational tone of voice. The credibility of your effort is very much established by the professionalism of your presentation. PowerPoint slides and visuals are required.

Assuming that your team will be required to make a formal presentation, the following suggestions will help you prepare for the task.

- Use the format of the EPR as your template for the final presentation.

- Create visual slides that not only synthesize ALL of the information, but also present the information in a visually pleasing manner. Do not use cute or goofy visuals; no team photo. Remember, this is about the project design. Most importantly, do not place too much information on each slide making it impossible for the viewer to read and understand the information. Be succinct but clear.

- In addition to the required technical information, integrate images and graphics including professional drawings, CAD drawings, sketches, equations, and free body diagrams.

- Carefully review the final slide presentation to make certain that it clearly organized, succinct yet thorough, no spelling or grammatical errors, consistent in format, and that all images are appropriate. It should have a sense of continuity—a consistent flow rather than a choppy sense that different people wrote various parts.

- Next, practice, practice, practice. Practice early; do not wait until the night before. Repeated practice helps you not only be prepared for the actual presentation, but it also helps you determine if the presentation flows well and if it will make sense to the listener. Setting up your phone or tablet to take a video of your practice sessions can help you check out your delivery as well as your posture, your voice projection, and eye contact. Remember to practice making a smooth transition between each team member's section.

- Plan your attire early. Imagine that you are interviewing for your first job. Dress professionally—a crisp neat appearance from head to toe. Make sure your shoes are clean and appropriate also!

- Some helpful ways to cope with pre-presentation jitters include the following.

    1. Remember the difference in distress and eustress.

       Distress occurs when you think you cannot do something. If you think you cannot do something, you probably cannot. You can reframe that kind of negative thinking by recognizing that the sensations you are experiencing in your body are the result of your recognition of the "demand" that making the presentation presents; that revved up feeling is actually your body's way of preparing you for the task and helping you be alert and ready to perform—just like it helps an athlete prior to a game or event.

    2. Replace negative self-talk with positive, self-supportive thoughts.

       Replace "I can't do this. I will be terrible" with "I know I can do this. I can learn a lot from this experience and get better at it in time."

    3. Use your imagination to visualize yourself being a calm successful presenter.

       World-class athletes regularly use visualization to prepare for competition. According to research involving brain imagery, visualization works because neurons in our brains interpret imagery as equivalent to a real-life action. When we visualize an act, the brain generates an impulse that tells our neurons to "perform" the movement.

    4. Take several long slow breaths every time you feel a wave of fear. Slowing down the breath is very calming to both your body and your mind.

    5. When you practice, and especially during the entire presentation, ground yourself by standing tall with even weight in both feet.

12.4   EPILOG

In *Keep Going: The Art of Perseverance*, Native American writer Joseph M. Marshall III recounts the story of a young man who turned to his grandfather for comfort after the death of his father. The grandfather shares his perspective on life, the perseverance it requires, and the pleasure and pain of the journey. We started this text with reference to capstone design being the great journey. It likely will be filled with high and low moments, times when the successful ending will seem just around the corner and times when nothing seems to work and starting over again from ground zero seems the only path to choose. There will be times when you will turn to your teammates for support and other times when you never want to see their faces again. That is the journey of capstone design. Above all else, a successful conclusion will demand perseverance. It is a quality—perhaps a better word is "virtue"—that will serve you well as you move into your professional career. Our last words to you are this—KEEP GOING!

Bibliography

[1] ABET. (2020). http://abet.org

[2] Ali, M. (2015). Ruthless prioritization and pragmatism. hotjar.com 44

[3] Anastas, P. A. (2013). *Innovations in Green Chemistry and Green Engineering: Selected Entries from the Encyclopedia of Sustainability Science and Technology*. New York, Springer. 38

[4] Anderson, L. K. (2000). *A Taxonomy for Learning, Teaching, and Assessing: A Revision of Bloom's Taxonomy of Educational Objectives*. New York, Pearson. 4

[5] Anthony, S. (2013, April 1). What is transhumanism or what does it mean to be human? http://www.extremetech.com/extreme/152240-what-is-transhumanism-orwhat-does-it-mean-to-be-human 29

[6] Aurelius, M. (2020). Meditations. Matosinyos.

[7] Baillie, C. (2012). *Heterotopia: Alternative Pathways to Social Justice*. John Hunt Publishing. 47

[8] Barry, M. (2012, March 31). Sustainability in ancient Rome. *Sustainability Institute*, University of New Hampshire. https://sustainableunh.unh.edu/blog/2012/03/sustainability-ancient-rome#::text=Sustainability%20in%20Ancient%20Rome.%20It%20will%20cover%20the,and%20supplying%20water%20to%20the%20empire%20with%20aqueducts 36

[9] Bellis, M. (2020, January 21). History of steam engines. *ThoughtCo*. https://www.thoughtco.com/history-of-steam-engines-4072565 25

[10] Berne, E. (1996). *Games People Play: The Basic Handbook of Transactional Analysis*. New York, Ballantine Books. 108

[11] Bjerklie, D. (1998, January 1). *The Art of Renaissance Engineering*. MIT Technology. https://www.technologyreview.com/1998/01/01/237121/the-art-of-renaissanceengineering/ 21

[12] Bobel, M. (2019). Wolf image gallery. *Animals: How Stuff Works*. https://animals.howstuffworks.com/mammals/wolf-pictures.htm 50

[13] Brand, S. (1971). *The Last Whole Earth Catalog*. New York, Random House. 97

[14] Braungart, M. A. (2002). *Cradle to Cradle: Remaking the Way We Make Things*. Northpoint Press. 56

[15] Buescher Jr., C. P. (n.d.). *History of Environmental Engineering*. Washington University in Saint Louis. https://eece.wustl.edu/eeceatwashu/about/Pages/environmental-engineeringhistory.aspx 36

[16] Campbell, J. (2014). *The Heroes Journey*. New York, New World Publishing. 1

[17] Carson, R. (2008). *Silent Spring*. New York, Penguin Books in Association with Hamish Hami. 36

[18] Catalano, G. (2019). Beyond traditional engineering. *ESJP*, Hankins. 40

[19] Catalano, G. D. (2014). *Engineering Ethics: Peace, Justice, and the Earth*, 2nd ed., *Synthesis Lectures on Engineers, Technology and Society*. San Rafael, CA, Morgan & Claypool. 92

[20] CATWOE Analysis. (n.d.). Free management books. http://www.free-management-ebooks.com/news/catwoe-analysis/#:~:text=%20CATWOE%20Analysis%20%201%20A%20Holistic%20Approach.,are%20involved%20in%20the%20situation%20and...%20More%20 69

[21] Code of Ethics. (2020). National society of professional engineers. https://www.nspe.org/resources/ethics/code-ethics 34

[22] Copernicus, N. (1995). *On the Revolutions of Heavenly Spheres*. New York, Prometheus Press. 23

[23] Core Values Clarification Exercise. (2016). University of Wisconsin Farm Extension. https://fyi.extension.wisc.edu/farmsuccession/files/2016/11/Core-Values-Exercise_Integrity_Consulting_Services.pdf 47

[24] Covey, S. (2013). *The 7 Habits of Highly Successful People*. New York, Simone & Schuster. 61, 109

[25] DaVinci Inventions—Inspired Engineering. (2014, November 4). *CNN*. https://www.cnn.com/2011/11/04/living/da-vinci-inventions/index.html 21

[26] Davis, J. A. (Director). (2016). The World's Columbian Exposition, NBC Series Timeless [Motion Picture]. 26

[27] DeGrasse, N. T. (2020). Neil deGrasse Tyson. Hayden Planetarium. https://www.haydenplanetarium.org/tyson/

[28] D'Olimpio, L. (2016, 2 June). The trolley dilemma: Would you kill one person to save five? *The Conversation.* https://theconversation.com/the-trolley-dilemma-would-you-kill-oneperson-to-save-five-57111 86

[29] Dudley, A. (2016). Squintin', lookin', doin'. http://amosdudley.com/weblog/SLO-Camera 28

[30] Dylan, B. (1964). The times they are a changin'. *Genius.* 33

[31] Dym, C. A. (2008). *Engineering Design: A Project Based Introduction.* New York, Wiley. 67, 69, 74

[32] Egolf, D. A. (2001). *Forming, Storming, Norming, Performing.* Bloomington, IN, iUniverse. 44

[33] Engineering, Social Justice, and Peace. (2020). http://esjp.org/ 39

[34] Feffer, J. (2019, July 31). Lifeboat Earth. *Common Dreams.* https://www.commondreams.org/views/2019/07/31/lifeboat-earth 97

[35] Frank, C. (n.d.). *Wikipedia.* https://en.wikipedia.org/wiki/Frank_Clark_(politician) DOI: 10.1109/icdm.2008.80. 105

[36] Frischman, B. S. (2018). *Re-Engineering Humanity.* Cambridge, Cambridge University Press. 30

[37] Galilei, G. (2017). *Dialogues Concerning Two New Sciences.* New York, CreateSpace Independent Publishing Platform. 19, 23

[38] Gould, J. A. (1985). *Communications of the ACM,* 28(3):300–311. 78

[39] Huxley, A. (2010). *Brave New World.* New York, Harper. 29

[40] Information Learning Activity—results and analysis. (2013, November 4). *Smithinfosearch.* https://smithinfosearch.wordpress.com/ 5

[41] Jeffrey, A. P. (2020). Personal Communication. 97

[42] Johnson, L. (1989). *A Morally Deep World.* Cambridge, MA, Cambridge University Press. DOI: 10.2307/2219690. 92

[43] Koberg, D. (1981). *The All New Universal Traveler: A Soft-Systems Guide to Creativity, Problem-Solving, and the Process of Reaching Goals.* New York, Kaufman Press. 87

[44] Kuehl, R. (1999). *Design of Experiments: Statistical Principles of Research Design and Analysis.* London, Duxbury Press. 101

[45] Larson, R. (2004). *The Great Wheel*. New York, Bloomsbury. 27

[46] Ilgin, M. A. (2010). Environmentally conscious manufacturing and product recovery (ECMPRO): A review of the state of the art. *Journal of Environmental Management*, pages 563–591. DOI: 10.1016/j.jenvman.2009.09.037. 79

[47] Mades, N. (2018, July 10). Top 15 best civil engineering quotes that every engineer can follow and share. *Quality Assurance and Quality Control*. https://www.qualityengineersguide.com/top-15-best-civil-engineering-quotes-that-every-engineer-can-follow-and-share 59

[48] Mitcham, C. A. (2010). *Humanitarian Engineering*. San Rafael, CA, Morgan & Claypool. DOI: 10.2200/s00248ed1v01y201006ets012. 39

[49] Mulder, P. (2016, September 12). TRIZ method by Genrich Altshuller. *Toolshero*. https://www.toolshero.com/problem-solving/triz-method/ 62

[50] Myers, J. (2019, April 30). *On Isle Royale, High Moose Population is Damaging Forest*. Twin Cities Pioneer Press. 16

[51] National Academy of Engineers. (2017). NAE Grand Challengs for Engineers. Washington, DC, NAE. 11

[52] Newton, S. I. (2016). *The Principia: The Authoritative Translation and Guide: Mathematical Principles of Natural Philosophy*. Berkeley, University of California Press. DOI: 10.1525/9780520964815. 23

[53] NSPE Code of Ethics Preamble. (2020). *NSPE*. https://www.nspe.org/resources/ethics/code-ethics 85

[54] Palermo, E. (2014, March 19). Who invented the steam engine. *LiveScience*. https://www.livescience.com/44186-who-invented-the-steam-engine.html 25

[55] Pinto, V. (1998). *Gandhi's Vision and Values*. New Delhi, Sage Publications. www.swaraj.org/visionandvalues.htm 43

[56] Quenck, N. (2009). *Essentials of MBTI Assessment*. New York, Wiley. 45

[57] Ricklef, R. A. (2015). *The Economy of Nature*. New York, W.H. Freeman and Co. 15

[58] Rissignol. (2012, December 31). Healthy ecosystsem. *ProProfs Quizzes*. https://www.proprofs.com/quiz-school/story.php?title=healthy-ecosystems 14

[59] Robinson, S. K. (2014, September 16). TedTalkxLiverpool. *TedTalk*. https://www.youtube.com/watch?v=FLbXrNGVXfE 11

[60] Schweitzer, A. (n.d.). Albert Schweitzer quotes. *Positive Life Project*. https://positivelifeproject.com/albert-schweitzer-quotes/ 85

[61] Technical Revolution. (2010, November 16). http://utaoct10.blogspot.com/ 25

[62] The Eds. of Encyclopedia Britannica. (n.d.). Sir Richard Arkwright. *Encyclopedia Britannica*. https://www.britannica.com/biography/Richard-Arkwright 24

[63] Thoms Savery invented the steam engine. (n.d.). *Famous Inventors*. https://www.famousinventors.org/thomassavery#:~:text=Thomas%20Savery.%20Thomas%20Savery%20invented%20%E2%80%9CSteam%20engine%E2%80%9D.%20Thomas,a%20thorough%20education%20and%20was%20fond%20of%20mathematics%2C

[64] Thurow, L. A. (2009). 3-D visualization of turbulent jet, division of fluid dynamics. *APS*. https://www.aps.org/units/dfd/pressroom/gallery/2009/thurow09.cfm 28

[65] Tudorache, T. (2005, December). Employing ontologies for an improved development process in collaborative engineering. *ResearchGate.net*. https://www.researchgate.net/publication/234016777_Employing_Ontologies_for_an_Improved_Development_Process_in_Collaborative_Engineering/figures?lo=1 55

[66] Twain, M. (2004). *Letters from Earth*. New York, Harper Perennial Modern Classics; Perennial Classics Ed. 113

[67] Varisty Tutors. (2015, July 28). How to choose your capstone project. *Varsity Admissions: An Educational Blog*. https://www.varsitytutors.com/blog/how+to+choose+your+capstone+project#:~:text=%20How%20to%20Choose%20Your%20Capstone%20Project%20,topic%20that%20you%20would%20like%20to...%20More%20 5

[68] Vitello, P. (2011, August 10). Ray Anderson, businessman turned environmentalist, dies at 77. *New York Times*. 58

[69] Wei, C. (2010, June 8). The overcrowded lifeboat. *Food for Thought*. https://weifood.blogspot.com/2010/06/overcrowded-life-boat.html xiii, 90

[70] Wiggins, J. (1996). *The Five Factor Model of Personality*. New York, Guilford Press. 46

[71] Williams, B. (2020). T552. *Systems Thinking and Practice*. http://systems.open.ac.uk/materials/T552/ 69

[72] wordpress.com. (2013). The dome of the Florence Cathedral-Filippo Bruneschelli (architecture). *Renaissance Art*. https://kingc16.wordpress.com/2013/09/23/the-dome-of-the-florence-cathedral-filippo-bruneschelli-architecture/ 22

Authors' Biographies

GEORGE D. CATALANO

George Dominic Catalano retired as Professor Emeritus of Biomedical Engineering at Binghamton University in 2019. Dr. Catalano was appointed a SUNY Distinguished Service Professor in 2018 for his many contributions to the university, and to the engineering profession. He earned a Doctor of Philosophy and a Master of Science degrees in aerospace engineering from the University of Virginia and a Bachelor of Science degree also in aerospace engineering from Louisiana State University. Prior to his time at Binghamton University, Dr. Catalano served on the faculty of the United States Military Academy, Louisiana State University, Wright State University, and the United States Air Force Institute of Technology. In addition, he was selected as a Fulbright Scholar at the Polytechnic Institute in Turin, Italy and the Technical Institute in Erlangen, Germany. Dr. Catalano also served on active duty as an officer in the USAF from 1973-1981. His research interests include turbulent fluid mechanics, low- and high-speed aerodynamics, experimental methods in physics, engineering education, engineering design, and engineering and environmental ethics. Dr. Catalano has more than 200 technical and educational publications and is listed in the Philosopher's Index for his published works in animal rights.

KAREN C. CATALANO

Karen Coltharp Catalano is a college counselor and instructor who specializes in teaching college students how to maximize their learning potential in highly competitive settings. Most recently she served on the faculty at Binghamton University while directing a program for underrepresented students in engineering and the sciences. She also served on the faculty at the United States Military Academy in the Center for Enhanced Performance and as a learning specialist in the Learning Assistance Center at Louisiana State University. She earned a M.Ed. in Counseling Psychology in 1980 from Louisiana State University and a B.S. from the University of Texas at Austin in 1972. Karen is a certified Svaroopa Yoga instructor and has taught yoga for over 20 years.

Printed in the United States
by Baker & Taylor Publisher Services